Public Economy and Urban Governance in China

Series Editor

Baijun Wu, East China University of Science and Technology, Shanghai, China

This book series aims to promote the research in innovation-driven development of public economy and urban governance in China, and ultimately the social governance model based on collaboration, participation, and common interests enhancing the foresight, precision, and efficiency of urban governance. Focusing on the urban governance of Chinese top megacities, this book series combines economics and sociology and explores a new way to solve problems of social governance concerning urban public goods supply mechanism, innovative models of social governance, as well as critical urban development issues like public safety, infrastructure and environmental pollution.

Jibo Yang

Effective Supply Mechanism and Path of Environmental Public Goods in China

ECUST PRESS

Springer

Jibo Yang
School of Business
East China University of Science
and Technology
Shanghai, China

Project Funded by Shanghai Universities Serving National Major Strategic Publishing Program and Series of Studies on Public Economy and Urban Governance

ISSN 2948-1872 ISSN 2948-1880 (electronic)
Public Economy and Urban Governance in China
ISBN 978-981-99-2938-2 ISBN 978-981-99-2939-9 (eBook)
https://doi.org/10.1007/978-981-99-2939-9

Jointly published with East China University of Science and Technology Press Co., Ltd.
The print edition is not for sale in China (Mainland). Customers from China (Mainland) please order the print book from: East China University of Science and Technology Press Co., Ltd.

This Springer imprint is published by the registered company Springer Nature Singapore Pte Ltd.
The registered company address is: 152 Beach Road, #21-01/04 Gateway East, Singapore 189721, Singapore

General Foreword

After more than 40 years of rapid development, Chinese economic society has shown some new characteristics in recent years, which mainly refers to the changes in ways of present economic development. This change in economic development after a certain stage is partly similar to the history of developed countries, but more reflects Chinese characteristics.

In terms of economic growth and structural adjustments, a remarkable feature is that the contribution rate of public goods sectors to national economic growth continues to rise. Especially after the global financial crisis in 2008, in order to resist the economic recession and maintain a high-speed economic growth and social stability, the Chinese government has continuously invested heavily in public infrastructure construction. At the same time, under the background of urbanization and population system reform, in order to meet people's increasing needs for social welfare, the government has increased its investment in public services including education, health, safety and social security, which makes the supply of economic and social public goods an important part of the total social supply. Even in some years, the investments and supply of public goods have become decisive factors for economic growth.

In terms of changes in social development and structure, although the overall social formation and power structure hasn't changed a lot, the social forces based on modern economic activities, such as new immigrants, middle-income class and private capital class, are increasing. And their influence on economic and social development is also increasing. These new social forces are subtly changing China's social structure and profoundly affecting the pattern of social interests. At the same time, in the course of people's increasingly marketized and socialized production and lifestyle, great changes have been witnessed in the ideology of grass-roots group, social needs and demands, which adds difficulties to how to accurately know people's preferences and behavior changes in social governance and challenges how to effectively implement public services. In addition, with the changes of production mode and the expansion of economic activity space, the social mobility is growing, which has gone far beyond the reach of administrative regions. This has brought great challenges to the traditional administrative management system and the top-down governance mechanism and put forward new requirements in modern political construction and economic construction for how to reconstruct the social governance structure at basic level and governance mechanism.

At present, it is difficult for us to form a definite judgment on the future development due to the changes and transformation of social economy. However, if we don't consider the factors of political system changes, we may be able to clearly find some laws of the evolution of social economy from that of social economy in various countries. Throughout the history of development in human society, changes in technology and production modes and market expansion provide economic security for people to get rid of the constraint of natural environment so as to pursue a prosperous life. However, the development of private product sectors can hardly ensure that every member of society enjoys a fair and safe development environment. The development of public goods sectors, whether under the planned economy system or the market economy system, the supply of public goods and public services bring social welfare promotion to the members of society. Therefore, the functional characteristics and social functions of the above two departments require a country to rationally allocate resources to the private product sectors and the public product sectors in the process of economic development; otherwise it will lead to a disharmonious relationship between economy and society. At the same time, in the development in private product sectors in many countries, as social wealth rises, the environmental damage, the gap between the rich and the poor and social injustice are becoming serious. Moreover, the development of public goods has also seen some problems such as resource mismatch, low supply efficiency, bureaucratic privilege and administrative monopoly in the course of improving social welfare level and safety guarantee. Therefore, people are always looking for appropriate systems to restrain the excessive competition, negative externalities and widening gap between the rich and the poor among private product sectors and also trying to improve the supply efficiency and social benefits in the development of the public goods sectors through systems.

From the trend of economic development in other countries, as the economy and personal income grows, the function and status of public goods sectors in economic and social development are gradually enhanced, and its asset scale and proportion in total social supply are also increasing, which reflect an improving state and trend of social economy. Referring to the law of development of public goods sector in other countries, the development of Chinese public goods sectors after reform and opening up also reflects the regularity of social and economic changes. However, we must realize that in the radically changing economic development, not only the growth of the private goods sectors will be affected by the changes of technology, market and system, but also the development of the public goods sectors will face many new challenges, which are mainly manifested in the following aspects:

First, how to rationally and effectively allocate social economic resources to private goods sector and public goods sectors. It is necessary to grasp the objective needs of economy and society in an all-round way and allocate resources with an effective mechanism, so as to realize the balance between the supply of two sectors and the needs of individuals and society and the balance between the economic growth and the improvement of social welfare to basically achieve the two goals of economic efficiency and social equity in development.

Second, how to effectively supply public goods, especially how to theoretically and practically solve the problem of truly showing people's demand preference and adopt appropriate public choice methods based on China's national conditions, so as to let the public participate in the selection and decision-making system of public goods projects. This problem involves a country's supply system and mechanism of public goods, as well as political, technical, market, social and cultural factors. However, its main problem lies in the arrangement of political and economic system and the design of mechanism. How to break the original "omnipotent government" and truly take the needs of residents as its core. And the government can analyze every situation, delegate its power to lower levels and release spaces for the development of social organizations and their performance of social governance functions. At the same time, the government should also reform its existing government-led public goods supply mechanism through institutional arrangements and political procedures to explore the multi-subject cooperation of public goods supply mechanism suitable for Chinese national conditions.

Third, how to build a social governance paradigm based on collaboration, participation and common interests and foster a community of social governance in which everyone fulfills their responsibilities and shares in the benefits. And how to establish a social governance system that integrates government, social organizations and the public and how to stimulate people's enthusiasm for participating in community's governance and improve the ability and level of participants in community affairs.

Fourth, how to innovate governance patterns in lower levels. This requires grass-roots governance to become the cornerstone and focus of national governance and the basic stage of modernization of China's system and capacity for governance. At the same time, through institutional arrangements and legal means the government should eliminate the "formalism" phenomenon of grass-roots governance, establish a governance mode that adapts to multiple governance subjects, to get rid of the administrative dilemma of community institutions and their affairs, explore the de-administrative reform path of grass-roots organizations in community and build the operation system and mechanism of grass-roots organizations in community in resource raising, public service provision and public affairs decision-making.

Fifth, how to innovate the production and supply of public goods, improve the efficient of public service and enhance the performance of grass-roots governance in community based on new technologies such as the internet, big data and artificial intelligence. This requires constantly improving laws and regulations, so that public service departments can legally use information technology to obtain the real needs of social members for public goods and our service, as well as provide sufficient information for public decision-making, technical means for public service supervision and performance evaluation, so as to achieve the goal of effective allocation of resources and supply.

Based on the above-mentioned thinking and understanding of the long-term development and structural adjustments in China's social economy, the Institute of Advanced Social Sciences of East China University of Science and Technology has set up a research team, which has established the research purpose of exploring the issues of public economy and social governance in China's economic and social development. It is hoped to have a subtle observation and partial analysis of China's social and economic development based on the social grass-roots level and micro-economic level under the grand paradigm of social and economic changes. In recent years, a group of young teachers of public economics, sociology and public management in East China University of Science and Technology have focused on public economy and grass-roots governance and undertaken many international cooperation and state-funded projects. They have carried out a series of studies around three major themes: First, they have studied the basic theory of development in public goods sectors and China's practice, especially through the integration of interdisciplinary research and multi-research paradigm, to explore the structural relationship and internal mechanism between public economy and social governance in the process of social development; the second is to study new modes and forms of public economy and social governance with new technologies including the internet, big data and artificial intelligence; the third is to carry out social investigation, field experiment and empirical analysis in public infrastructure, public service and community governance, trying to find some useful practical experiences with Chinese characteristics, so as to review and revise the classical theoretical views and expand the scope of theoretical research. This series is a preliminary summary of our previous research work, and each fascicle also roughly reflects the research topics in these aspects.

The contents of this series mainly involve the fields of public economic, social governance and social policies. No matter which discipline they belong to, most of the studies are based on the theories of public economics and public management and discuss the scientific problems in public economy and social governance in an interdisciplinary way; in view of policies, we also study it from the perspective of various disciplines. Four books in the series can be classified into public economics. Among them, *"Effective Supply Mechanism of Urban Public Goods in China"* sorts through the supply theory and research literature of public goods, discusses the mechanism of Chinese public goods supply and its gains and losses in reform and focuses on the empirical analysis of investment, financing, production and distribution in the fields of urban public infrastructure and public service; *"Urban Infrastructure and Economic Development in China"* summarizes the achievements and challenges of Chinese society in infrastructure construction in the past 70 years and reveals the internal mechanism of infrastructure affecting regional economic development from the theoretical perspectives of fiscal decentralization and comparative institutional analysis; using econometric analysis, this book explores the influence of Chinese infrastructure construction on the integration of the market, the open development and the improvement of people's livelihood and puts forward the theory and realization path of infrastructure supply model innovation from the perspective of political economy theory; *"Effective Supply Mechanism and Path of Environmental Public Goods in China"* studies the supply mechanism of various environmental public goods, focusing on the function and behavior pattern of household sector in environmental governance that people pay less attention to; taking the household sector as the supplying main body, the intergenerational factors are included in the analytical framework, and the theoretical model of effective supply of environmental public goods is constructed based on three valuation methods of environmental public goods. Interestingly, *"Multi-dimensional Collaborative Governance of Urban Sharing Platforms"* takes sharing economic as the research target and discusses the widespread publicity of private products. The book comprehensively sorts through the development mode, governance experience and action mode of urban sharing platforms at home and abroad and puts forward the concept of multi-collaborative governance of sharing platform with self-governance priority, government supervision and social assistance. This research breaks through the traditional restriction of supply of urban public goods and makes a pioneering research on the revolutionary changes of public consumption and production mode of private goods in the new technology environment. Generally speaking, these books take infrastructure, environmental protection, pollution control and other major urban public goods and public service as research targets, analyze the characteristics and internal operation mechanism of Chinese urban public goods supply and put forward plans and policies for the reform of effective supply system of public goods. There are two books in the series that study the urban social governance. *"Urban Migration and Public Governance in China"* investigates urban governance in the process of urbanization and studies how to transform immigration control into immigration service and improve the level of immigration governance so as to enable immigrants to integrate into cities as soon as possible; targeting at the reality of frequent accidents in urban public security, *"Urban Public Security Risk Prevention and Control System"* systematically analyzes the sources and types of urban public security problems. On the basis of deeply analyzing the causes of urban security risks, it discusses the effective internal mechanism of prevention and control of security risks and constructs the prevention and control system of urban security risks.

After the publication of this series in Chinese, it has attracted wide attention from academia and society, especially how to coordinate the development between the public sector and the private sector, how to establish a multiple public goods supply mechanism with social organization and public participation and how to establish an effective grass-roots social governance mechanism. Now the series is published in English, which can make more English readers understand China's great changes in the public economic sector and social governance field since reform and opening up and know Chinese scholars' understanding of realistic problems of public economy and social governance. At the same time, it is expected that scholars from all over the world will pay more attention to the reform process of Chinese public sectors and promote international academic exchanges in this field.

The publication of this series has been supported by the National Publishing Fund and the "University Serving National Major Strategic Publishing Project" combined with Shanghai Culture and Education Project. Thanks to the sincere support of East China University of Science and Technology Press and the hard work of editors, we can get financial support and high-quality publishing! Thanks to the support of

Springer Nature and the high-quality work of editing and translation, this series can be published in English.

Shanghai, China
May 2023

Baijun Wu

Foreword

Jibo Yang has researched China's economic development for a long time. Initially, she researched the urban real estate market, later shifted her area of research to public goods supply mechanisms, public education, family relations and conducted lots of researches on intergenerational public goods and family economic relations. In the period of studying for a Ph.D., she focused on researching the city environment based on her preliminary research on public economy. In this area, she has a background in environmental science and engineering and can accurately grasp the technical issues of environment. This has been a great help for her environmental economic research.

In her over three years of research, she carried out field surveys and collected lots of materials in connection with environmental pollution and governance. Since the beginning, this study has tried to take the environment as an example to crack some difficult problems in the supply and demand of public goods, e.g., the measurement of the public goods supply mechanism and the effectiveness thereof, the measurement of the revelation of residents' real demands and preferences for the environment, and the externality characteristic of environmental public goods. In particular, she introduced intergenerational variables into her study to analyze the effect of intergenerational externality. This is a common problem which is seldom studied and difficult to study among the effects of public goods. In the study of public goods, much of the focus is on the spatial externality. However, due to a lack of a theoretical analysis method, it is difficult to obtain data available for measuring environmental intergenerational effects, and there is little progress made in this aspect. Therefore, this book has conducted a trial study (choosing to research the intergenerational effects of environmental public goods from the perspective of the household sector) and made achievements with theoretical contributions.

Focusing on the aforesaid themes, this book has certain theoretical innovation or application value in the following aspects.

First, in the study of the environmental governance supply mechanism, the household sector is taken as the supply subject, with a focus on demonstrating the feasibility of residents shifting from the demand side of clean environment to the supply side. Based on related data from the Chinese General Social Survey (CGSS), this part analyzes the degree of residents' concern for environmental pollution, their willingness to participate in environmental governance and action therein. Thus, its conclusion is credible in this regard. With the progress of society and the increase of residents' income, the household sector is supposed to have an increasingly larger influence on environmental governance. On the basis of statistical analyses, this book also employs three methods of public goods valuation—the revealed preference method, the stated preference method and the happiness measurement method to analyze the revelation of residents' environmental demands and preferences. Through a comparative analysis, this book concludes that the happiness measurement method is a preferable method of measuring residents' willingness to pay for environmental protection, and draws an empirical conclusion that residents have their intrinsic demand motivation for participating in environmental governance. It is of great importance that it theoretically demonstrates the internality of the household sector and residents shifting to the supply side of public goods, and lays an academic and theoretical foundation for the multi-subject environmental governance system where the government, enterprises, and residents reach tripartite cooperation.

Second, in the literature study on the supply of environmental public goods, this book reviews such theories as intergenerational externality, altruism, public goods supply mechanism, public choice, contract, and motivation, analyzes the intergenerational externality of environmental public goods, and proposes that the spillover of intergenerational costs or benefits will give rise to cost–benefit asymmetry between the "Present Generation" and "Future Generations." Starting with the rational man hypothesis, this study holds that the present generation makes decisions on their own side, with little consideration for the interests of future generations. As a result, compared to "Backward Intergenerational Public Goods" which bring benefits to the present generation or provide them old-age security, "Forward Intergenerational Public Goods" to the benefit of future generations are in seriously short supply, which is one of the reasons why intergenerational public goods like environmental protection are in short supply. Therefore, the author considers that it is necessary to research and explore a feasible way to promote the intergenerational allocation of resources in a rational and efficient manner, improve the supply level of intergenerational public goods, and pursue the intergenerational equity of public goods supply. For this end, guided by the altruistic motive proposed by Becker, this book builds an overlapping generations (OLG) model for residents' participation in the supply of environmental public goods and provides a new research idea for increasing the supply of "Forward Intergenerational Public Goods" beneficial to future generations. This result deserves a further discussion among theorists. We hope that it could arouse the attention of the government sector.

Third, this book anatomizes China's environmental status, analyzes the sources of pollution with technical data, and finds that in addition to the discharge of industrial wastewater, waste gas, and so forth, which is the principal factor, domestic wastewater, household waste, etc., are increasingly becoming important factors of environmental pollution. However, these household sector factors are neglected in the existing environmental governance policies, environmental regulations, and supply mechanism of environmental public goods. Therefore, this book probes into how to guide residents to participate in environmental governance and become the supply subject of environmental public goods. In fact, with the development of society and the improvement of residents' environmental awareness, there is huge potential in this regard. The nationwide household waste treatment campaign launched in recent years has made a big start. The author should carry out a further study on this basis to make a contribution to enabling the environmental governance action dominated by the household sector to be the real subject of the governance system.

Fourth, guided by the polycentric theory and self-governance proposed by Elinor Ostrom, this book designs the multi-subject supply mechanism of environmental public goods led by the government, probes into and compares two supply modes, i.e. supply by the government and joint supply by the government, enterprises, and residents, optimizes the supply subject structure and its relations, and enriches the theoretical model for mechanism selection. Moreover, based on the overlapping generations (OLG) model introduced by Peter Diamond, this book builds a theoretical model for effective supply of environmental public goods—mechanism for government-led supply of environmental public goods by residents and enterprises. With the help of model deducing, this book proves that residents' participation in environmental governance can bring the following benefits on certain conditions: improving environmental quality, promoting economic growth, and benefiting the healthy growth of future generations. In the subsequent empirical study, taking air pollution control as an example, this book probes into residents' willingness to pay for environmental protection, as well as their individual characteristics such as educational attainment, income level, and having offspring or not, which are factors influencing their participation in environmental governance; introduces such cases as the "Five Water Co-governance" in Zhejiang (sewage treatment, flood control, waterlogging draining, water supply guarantee, and water saving) and the waste sorting in Shanghai which have achieved initial results, and expounds the feasibility and strategy of residents' participation in environmental governance. In the case study, the author employed the happiness measurement method and the supply mechanism of environmental public goods and put forward practical governance methods thereof.

Fifth, on the basis of analyzing the institutional root causing the misallocation of China's environmental public goods, this book puts forward a series of strategies and suggestions for the effective supply thereof, e.g., the credible commitment of the "Long-term Government," the cross-generational subsidy policy in the "Intergenerational Contract," the policy that gives both penalties and subsidies to polluting enterprises, the establishment of a "Environmental Protection Fund Company" to manage environmental governance payments and revenue. Most of these suggestions are proposed upon a feasibility assessment based on actual conditions and play a role in supplementing and improving the existing policy and mechanism reform. As this book advocates, only the coordination of each party's responsibilities and interests can motivate residents and enterprises to protect the environment, solve the inefficiency and "free-rider" problems of environmental public goods supply, and ensure the effective implementation of a mechanism for government-led joint supply of environmental public goods by enterprises and residents.

Jibo Yang's study shows that China's rule of law system, industrial foundation, governance mechanism, and public awareness regarding environmental governance remain weak. However, compared to other countries, in terms of the investment in environmental governance and the importance attached thereto, China has outperformed the vast majority of countries in similar stages of economic development and achieved significant governance performance in recent years. Now, the key task is to establish a long-term and stable governance system and mechanism and carry forward the environmental protection culture rooted in the society and the public.

As regards the supply of public goods in environmental governance, China's system foundation has certain particularities. Government dominance and administrative compulsion appear significant, but a governance system or mechanism based on the principles of market transactions and social self-governance is far from being established. In China's environmental governance, administrative compulsion and campaign-style governance actions have large direct costs and indirect losses. Ultimately, it is difficult to achieve long-term goals in reliance upon external compulsory regulation. Moreover, due to the repeated changes in governance requirements, standards and policies, enterprises and residents tend to show low response thereto. Their avoidance and gaming behaviors may greatly weaken the performance of governance in general. There are many reasons for this occurrence, and one of the important reasons is the lack of an appropriate governance mechanism based on the balance of interests.

The practice at home and abroad suggests that it is difficult to realize the effect of environmental public goods supply and the demand satisfaction thereof, without regard to the cost-benefit relations between stakeholders and the competitive relations between behavior subjects in the process of environmental governance. It is gratifying that with the progress of technology and the enriching of financial instruments and market transaction means, the cost of environmental governance will continue to decline, with more certain governance performance. In this regard, this study bears this out and embodies the value of this book.

Shanghai, China
November 2019

Baijun Wu

Preface

The report of the 19th CPC National Congress expressly points out that the principal contradiction in Chinese society has evolved into one between unbalanced and inadequate development and the people's ever-growing needs for a better life. General Secretary Xi Jinping proposed, "Ensuring harmony between human and nature, we must realize that lucid waters and lush mountains are invaluable assets and act on this understanding."

However in these years, with the rapid development of economy, the supply contradiction of environmental public goods has become increasingly prominent, and the problems emerging in the supply process have become increasingly severe, especially air, water, and waste pollution. Such pollution has seriously damaged our living environment, disturbed our lives, and hindered our progress from material to spiritual life. How do we balance the relations between development and the people's pursuit of a better life, and effectively supply environmental public goods? This is a core problem to be discussed in this book.

At present, the supply modes of public goods comprise government supply, market supply, voluntary supply, and mixed supply. The "free-rider" problem of public goods can be extended to the conclusions that government governance is the most efficient, and the government may fail. Then, the market mechanism can be introduced into the supply of public goods, followed by a market-oriented and socialized multi-subject governance mode. Which operation mechanism can reveal residents' preference for public services to the maximum extent? Which mechanism is more effective for China's environmental pollution control? These are important social concerns at the present time.

Considering that the research on the supply of environmental public goods is complicated, and connected with such theories as intergenerational externality, altruism, public goods supply mechanism, public choice, contract, and motivation, this book starts from the environmental status to analyze the nature of problems, and reveals that residents, as the demand side of quality environment, is also one of the pollution sources in fact. Focusing on residents, this book extends the application of the public goods theory to environmental governance, analyzes residents' effective demand revelation and how to avoid "free-riding" in the supply process in terms of the spatial and intergenerational externalities of environmental public goods, sets forth that the happiness measurement method is capable of measuring residents' willingness to pay for environmental protection to some extent, and further uses a theoretical model to put forward the mechanism for government-led joint supply of environmental public goods by residents and enterprises, thereby providing a certain theoretical basis for the effective supply of environmental public goods. In addition, this book strives to reflect the latest trends, new ideas, and new knowledge of environmental public goods supply by providing classic cases regarding residents' participation in environmental supply, i.e. the waste sorting in Shanghai and the "Five Water Co-governance" in Zhejiang.

Most of the content hereof comes from the author's research results of recent years. Wu Zhongjie, Sun Yanping, Du Juan, et al. successively participated in this study. Part of the content of Chap. 5 was written by Sun Yanping; some information of Chap. 9 was provided by Du Juan, and written by Jibo Yang; the case study in Chap. 10 was mainly written by Wu Zhongjie; the remaining chapters were completed by the author.

It should be particularly noted that our study has been funded by the "Study of Supply Mechanisms of Urban Public Goods" (2017-01-07-00-02-E00008), a major project in the Scientific Research and Innovation Program of the Shanghai Education Commission, as well as the "Shanghai Universities Serving National Major Strategic Publishing Program" Combining Culture and Education, the Major Courses of Shanghai, the ECUST Education and Teaching Method Reform and Research Project, the Shanghai First-class Undergraduate Construction Leading Plan, etc. Besides, most of the survey data comes from the CGSS project team. Here I would like to express my gratitude to them.

The effective supply of environmental public goods is a field under rapid development and exploration. There may be some errors or improprieties in the description and adduction of some contents. We sincerely welcome any criticisms and corrections from all walks of life.

May 2020

Jibo Yang
East China University of Science
and Technology Press
Shanghai, China

Acknowledgements

Writing a book is like conceiving a child. Finally, it is time to write my acknowledgements. Looking back at this process, I have mixed feelings. It is hard to explain in a few words. There are so many people who have supported and encouraged me, so many people I would like to thank.

First, I would like to extend my sincerest gratitude to my doctoral supervisor Prof. Wu Baijun., who has encouraged me to continue to explore what I want to research, guided me step by step on how to absorb essence from classic documents, and how to open my mind to combine interests and research since the topic selection of this book. His great foresight and composed analysis have led me to gradually grow in the academic exploration day by day; in the writing process of this book, he had fully affirmed and encouraged my effort, thereby enabling me to have confidence to overcome difficulties and return to a smooth road.

Besides, for the completion of this book, I would also like to express my thanks to the CGSS project team of the National Survey Research Center at Renmin University of China for providing me data assistance. Nevertheless, the author is solely responsible for the book content.

My appreciation also goes to my colleagues and friends, Mr. Pan Chunyang, Ms. Chen Yajing, Mr. Zhao Yan, Mr. Li Youping, Mr. Jiang Shiping, and Mr. Yang Jianxia et al. who have tirelessly listened to my exposition, and pointed out the deficiencies of my study. Some have provided me cutting-edge literature and kept discussing it with me. In short, it is a blessing to meet such like-minded friends in my life. I am deeply obliged.

In the end, I owe many thanks to my family. They have always made arrangements for all trifles in life, allowing me to be concentrated on writing. I am grateful to my husband for tolerating my bad temper in my writing period and acting as my driver in my interview and field survey process. He propitiated and comforted me when I was depressed; mollified and helped me when I met difficulties; enlightened and encouraged me when I was at a loss. Additionally, I wish to thank my daughter for cheering me up when I was stranded, "Mom, I believe you can successfully accomplish it. Come on! Let's work together." Their support has given me a great deal of encouragement and confidence. Family affection lasts forever. Thank you for being always on my side.

There are many other friends and teachers who have helped and supported me. Though it is inconvenient to mention all of them here, I would still like to express my sincere thanks.

About This Book

This book mainly focuses on the supply of environmental public goods. Starting from the spatial and intergenerational externalities thereof, it teases apart and reviews relevant concepts and theory schools, and probes into how to promote residents' participation in designing an effective supply mechanism of environmental public goods under the guidance of the government. The innovation of this book lies in the building of an overlapping generations (OLG) model for residents' participation in the supply of environmental public goods, and the provision of a new research idea for the long-term serious undersupply of "Forward Intergenerational Public Goods" beneficial to future generations in reality. In addition, the author used the happiness measurement method to measure the feasibility of residents' participation in the supply of environmental public goods, carried out a comparative study on two supply modes of environmental public goods, i.e. supply by the government and joint supply by the government, residents and enterprises, and proposed feasible strategies and suggestions on a theoretical basis.

This book can be used as not only a reference for professionals of the government, universities, and research institutions engaged in environmental governance, but also a reference book for related majors of universities.

Contents

Chapter 1
Introduction

With the rapid growth of the Chinese economy, insufficient domestic demand, insignificant technological progress, high environmental costs and other latent macroeconomic risks have increasingly emerged in the process of economic development, and air pollution, water pollution, waste pollution and other problems accumulating on a long-term basis have become particularly apparent. Environmental pollution has increasingly become a socioeconomic difficulty besetting sustainable development. How to take effective measures for environmental governance has become a focus of national attention. Can residents as the demand side of quality environment become the supply side as well? How do we bring social power into environmental governance? What responsibility should the government assume? All these are problems to be thought about and solved in the process of environmental pollution improvement.

1.1 Research Background and Significance

1.1.1 Research Background

Environmental public goods refer to various environmental goods and services, including air, water, forests, living environment and other natural environments, as well as transportation, artificial protection forests, systems and other artificial environmental services. They can directly or indirectly influence the life and development of mankind [1]. Through an analysis on the two characteristics of public goods—non-rivalry and non-excludability, it can be inferred that fresh air, high-quality water resources, and good living environment are typical public goods as they all have non-rivalry and non-excludability. In 2013, General Secretary Xi Jinping stressed on an inspection tour in Hainan, "A good ecological environment is the fairest public

J. Yang, *Effective Supply Mechanism and Path of Environmental Public Goods in China*, Public Economy and Urban Governance in China,
https://doi.org/10.1007/978-981-99-2939-9_1

product and the most accessible welfare for the people". Such environmental public goods as environmental quality, environmental policies and public environmental facilities have a far-reaching influence on our lives.

However in reality, we are not only affected by haze, sand storm and other severe weather, but also nagged by such problems as water pollution, noise, food, industrial and household waste pollution, insufficient green space, destruction of forest vegetation, degradation of farmland quality, shortage of fresh water, desertification and reduction of animals and plants. Environmental pollution not only greatly lowers our work efficiency, human capital and residents' happiness, but also severely affects our health, leads to the rise of the death rate of lung cancer and the incidence of cardiovascular diseases, and shortens residents' life expectancy [2–4]. Chen et al. [5] found that every 100 $\mu g/m^3$ increase in the total suspended particulate (TSP) concentration could reduce residents' life expectancy by three years. Researches indicate that [6] with every 1% increase in sulfur dioxide (SO_2) emissions, the number of people dying of respiratory illnesses and lung cancer will respectively increase by 0.055 and 0.005 among 10,000 people. The masses have made a strong appeal for improving the ecological environment. Although the government adopted a series of rationalized measures including closing, suspension, merging and shifting, as well as other measures to increase punishment, the effect was insignificant. The Outline of the "13th Five-Year Plan" presents the concept of "Innovative, Coordinated, Green, Open, and Shared" development. According to the report of the 19th CPC National Congress, as socialism with Chinese characteristics has entered a new era, the principal contradiction in Chinese society has evolved into one between unbalanced and inadequate development and the people's ever-growing needs for a better life. The people have raised higher requirements for material and cultural life, especially for high-quality environment. Thus, in the future socioeconomic development, how to build an effective supply mechanism of public goods that is open, sustainable and capable of continuously satisfying economic development and residents' needs has become a practical problem to be urgently solved.

Environmental governance has the typical characteristics of public goods. Its non-rivalry on consumption and non-excludability on benefits make it impossible to truly embody effective demands, difficult to fix a price and incapable of avoiding the "free-rider" problem in the supply process. If the value of environmental protection can be accurately assessed, and the price or tax that the people are willing to bear for curbing environmental pollution can be estimated, a great help will be provided for the effective supply of environmental public goods and the government's environmental governance. Additionally, environmental public goods are intergenerational public goods with obvious temporal characteristics. Intergenerational public goods (IPGs) are derived from the concept of intergenerational goods proposed by Sandler and Smith [7, 8], who have indicated that intergenerational goods are public goods shared across generations. The IPGs mentioned herein refer to the public goods objectively existing in the context of overlapping generations economy, i.e. goods shared and used by more than one generation. Compared to generational public goods, they have obvious intergenerational externalities [9, 10]. Intergenerational public goods include the following three factors [11]: (1) Temporality, with which both the inter-temporal investment and return of IPGs and the service life and utility thereof are associated; (2) stability, i.e. how to correct, maintain and continue the supply behavior of IPGs; (3) intergenerational "non-excludability and non-rivalry", i.e. the non-excludability and non-rivalry of public goods for over one generation of consumers [12, 13]. It follows that intergenerational public goods represent not only a generational issue relating directly to the present generation's life quality and development, but also an intergenerational issue concerning future generations' right to survive and immediate interests.

For a long time, there have been the phenomena of undersupply and misallocation in the supply of intergenerational public goods with obvious temporal characteristics in China, e.g. basic education, medical and health care, pure science and environmental protection. Starting from the rational man hypothesis, the present generation makes decisions on their own side, with little consideration for the interests of future generations. As a result, compared to the "Backward Intergenerational Public Goods" (e.g. old-age insurance) which provide old-age security for the present generation, the "Forward Intergenerational Public Goods"[1] (e.g. environmental protection, education) to the benefit of future generations are in seriously short supply. And the point is that intergenerational public goods are capable of improving the level of social welfare. Thus, researching and exploring a feasible way to promote the intergenerational allocation of resources in a rational and efficient manner, improving the supply level of intergenerational public goods, and pursuing intergenerational equity is a common topic for the entire human race. In consideration of China's "Fiscal Decentralization" and "Government Performance Assessment" status, we have further studied the root of misallocation of intergenerational public goods, and found that the government, motivated by the behavior of "being responsible to superiors", tend to provide intergenerational public goods (e.g. construction of infrastructure like roads and bridges) capable of promoting economic growth in a short term (or within a term of office), and might neglect those (e.g. education, environmental protection and other public services) capable of improving the people's livelihood, which are the essential conditions of boosting the long-term sustainable economic growth. What way should we choose to provide intergenerational public goods so as to ensure sustainable economic growth and maximize social welfare? This is an issue of concern in this book.

The above series of practical issues and theoretical problems are both connected with the effective supply mechanism of environmental public goods. In the existing political and economic system, the government sector has a strong motivation for attracting element agglomeration and promoting economic growth, but lacks mature experience in effectively allocating public resources, rationally providing environmental public goods and easing agglomerated negative externalities, and has no

[1] According to Antonio Rangel (2003), forward intergenerational public goods refer to public goods, e.g. public education and environmental protection, which will have an effect on future generations upon investment, with the present generations as the object of study. Backward intergenerational public goods refer to public goods which will have an effect on the previous generation upon investment, e.g. medical insurance and old-age security.

proper mechanisms or adequate incentives, resulting in the development dilemma of "Market Failure" and "Government Failure" coexisting.

In view of the above research background and fact judgments, this book intends to start from the perspective of residents' participation in environmental governance based on the characteristics of intergenerational externalities of environmental public goods, employ the overlapping generations model [14] introduced by Diamond, use the new progress in recent theoretical research for reference, attach full importance to the "free-riding" characteristic of such goods in the supply process, and further carry out empirical and theoretical researches on the basis of the preliminary quantitative analysis and empirical research, with a focus on exploring a public goods supply mechanism that is capable of effectively motivating residents to participate in environmental governance.

1.1.2 Research Significance

Considering the non-excludability, non-rivalry, non-divisibility in consumption utility and nonprofit nature of environmental public goods, the real demand thereof can be hardly accurately estimated, and the input cost and the return on investment are imbalanced. Because of these problems, the market mechanism is incapable of executing effective resource allocation as per the price system, which may easily give rise to a market failure. Scholars advocate the supply of public goods by the government, but the government's monopolistic practice and inevitable rent-seeking usually lead to the inefficient supply by the government, and may cause a government "failure". The voluntary supply of public goods by third-party social organizations seems ideal, but the inconsistency between their goals and social goals, and the diversity of types of environmental public goods may both lead to a failure in voluntary supply.

The theoretical research on the effective supply of environmental public goods still has numerous difficulties: how to create regulations so as to motivate consumers to reveal their true preferences; how to measure consumers' real willingness to pay for environmental protection and solve the charge problem; how to motivate residents as the demand side of clean environment to become the supply side of quality environment as well; how to solve the intergenerational equity in the provision and use of environmental public goods; how to develop a proper system to safeguard investors' rights and interests, etc. In particular, the supply of public goods is constrained by the basic socioeconomic system of a country, and the suitability and long-term utility of the supply mechanism may vary on different institutional conditions, thus the effective supply mechanism of environmental public goods must have institutional suitability.

This book carries out a study on the above theoretical and practical problems, and is of academic and practical research significance in the following aspects.

(1) This book anatomizes the regional differences, formation reasons and characteristics of three environmental public goods, respectively air pollution, water pollution and waste pollution, which are typical intergenerational public goods and residents' biggest concerns. This could help us to understand the current practical condition of allocation of China's environmental public goods. Besides, according to the types of environmental public goods, we have designed different supply mechanisms to solve the distortion of public goods demanders' preference revelation, identified the internal relations between marginal costs and benefits of private and public goods, put forward the theoretical conditions of achieving a balance between marginal social costs and marginal social benefits, further studied and proposed a modified theoretical model for the effective supply of public goods, and achieved theoretical innovation on public economics.
(2) Guided by the polycentric theory and self-governance proposed by Ostrom [15], this book probes into the thought of public governance fitting in with China's socioeconomic system, designs the multi-subject supply mechanism of public goods to theoretically optimize the supply subject structure and the relations thereof; discusses the effective supply mechanism of public goods constrained by China's political and economic system on the basis of the theory of collective action proposed by Olson [16] to theoretically enrich the theoretical model for mechanism selection; builds the "two-sector model of environmental public goods" (Diamond's OLG model) supplied by enterprises and the government under the compulsory intervention of a "visible hand", intends to achieve the equilibrium condition for supply of environmental public goods through a discussion on the gaming between enterprises and the government; further builds the "three-sector supply model of environmental public goods" (Diamond's OLG model with the addition of residents) covering residents, extends the supply to the joint supply by the government, enterprises and residents, and intends to maximize social welfare or achieve Pareto improvement while realizing the government-led supply efficiency of environmental public goods. As an extension of traditional research, it is of theoretical innovation value.
(3) This book introduces the method of measuring residents' willingness to pay for environmental protection—happiness measurement method (also known as life satisfaction measurement method), thereby solving the "free-rider" problem of public goods to some extent. In this study, we have not only measured residents' marginal willingness to pay for environmental protection, and scientifically expounded the feasibility of residents shifting from the demand side of quality environment to the supply side, but also could popularize this method in the field of public goods. It is of great theoretical innovation value and practical significance.

(4) In the existing practice on public goods supply, financial market instruments have been extensively used to introduce social capital. However, in the theoretical models of public economy concerning the fiscal decentralization and the regulation theory, the financing and operation style of public goods supply is single, and generally confined to the government sector, tax revenue, government debts, payment by consumers, and the like. Thus, it is difficult to interpret the public goods supply mechanism and investment and financing cases in the real economy, and fails to profoundly reveal the relations between supply subjects. This book will analyze the investment and financing mechanism, the production and operation mechanism, and the definition of contractual rights under the multi-supply mode in combination with the practical cases of environmental public goods, thereby giving an economically theoretical explanation for the multi-supply mechanism thereof.
(5) This study is conducive to improving the matching degree of social demand and supply of environmental public goods. Theoretically, the lack of a market transaction mechanism of public goods requires the government sector to keep up with the needs of residents and enterprises so as to provide various public goods in keeping with micro individual preferences. At the current stage, with the optimization and upgrading of the economic structure and the rapid development of society, the traditional supply mechanism can hardly meet the needs of industries and social groups. Therefore, this book reveals the preferences of such micro individuals as residents and enterprises by certain mechanism design, and brings them into the theoretical model for the supply mechanism as important parameters so as to improve the aforesaid matching degree.

It can be inferred that it is of great theoretical and practical significance to research environmental governance strategies from the perspective of residents' participation therein, based on the intergenerational characteristics of environmental public goods under the new normal.

1.2 Research Objectives, Framework and Ideas

1.2.1 Research Objectives

From the perspective of residents' participation in environmental governance in combination with the historical requirement for transformation of China's economic development mode, this study focuses on the intergenerational externalities of environmental public goods, with an aim to design an effective supply mechanism thereof so as to realize the supply efficiency of intergenerational environmental public goods with the improvement of social welfare, and explore a rational and effective mode and implementation path of environmental governance.

To achieve the above objectives, we have determined the following specific stage objectives:

(1) Based on a comparison of foreign and domestic researches, interpret the meaning and characteristics of intergenerational externalities of environmental public goods, and get an understanding of the current practical condition of allocation of China's environmental public goods by measuring the supply efficiency of three representative types thereof, i.e. air pollution, water pollution and waste pollution.
(2) Build Diamond's dynamic model based on Tiebout's [17] mechanism on the market-oriented condition. First, probe into the "two-sector supply model" where only enterprises and the government participate in pollution control, analyze Pareto optimality conditions for the supply of environmental public goods, and further introduce residents thereto; build the "three-sector supply model" where the government, enterprises and residents jointly participate in pollution control, dynamically analyze the supply of environmental public goods and local government acts, and intend to obtain the best way to realize the government-lead supply efficiency thereof.
(3) Compare the revealed preference method, the stated preference method and the happiness measurement method for environmental quality assessment, analyze the scientific methods for valuation on public goods, clarify the theoretical mechanism and comparative advantages of happiness measurement, specifically measure Chinese residents' marginal willingness to pay for environmental protection by the method, and further compare the differences in residents' marginal willingness to pay between countries.
(4) Explore the institutional root causing the misallocation of China's environmental public goods, identify the formation mechanism for Chinese residents' marginal willingness to pay lower than that of developed countries, and propose the government-led supply mechanism of environmental public goods with the joint participation of residents and enterprises on this basis, thereby providing a reference for government decision-making.

1.2.2 Research Framework and Ideas

Taking environmental public goods as the object of study, this book focuses on exploring the effective supply mechanism and path thereof. From the perspective of residents' participation in environmental governance, it starts with the characteristics of intergenerational externalities to build an OLG model, and to explore an incentive mechanism where the government, enterprises and residents jointly participate in environmental governance and the implementation measures thereof.

On the basis of literature review and research, this book conducts theoretical modeling in reliance upon the public choice theory, the principal-agent theory, the contract theory, as well as other theories and methods, with a focus on research problems. In this connection, this book first measures the production efficiency and allocation efficiency of environmental public goods to get an understanding of the performance of government inputs and outputs on environmental quality improvement; introduces Diamond's OLG model to theoretically build an equilibrium model (modified Diamond's OLG model) where enterprises and residents jointly participate in government-led environmental governance; further employs the happiness measurement method, theoretically probes into the feasibility of residents' participation in environmental governance, and empirically analyzes residents' marginal willingness to pay for environmental protection in different parts of China; explores the institutional root causing the misallocation of China's environmental public goods on this basis, identifying the reasons why Chinese residents' marginal willingness to pay is lower than that of developed countries, and further proposes the government-led supply mechanism of environmental public goods and the strategies thereof. The specific research ideas and framework are described as follows.

(1) The classic theories on environmental public goods with intergenerational externalities and the modern development thereof, the theoretical and empirical researches regarding the impact of intergenerational public goods on society and people's livelihood, as well as the institutional factor influencing the supply decision of environmental public goods, and the three valuation methods thereof are systematically reviewed. This is the literature foundation of this study.
(2) With a focus on reviewing the status of air, water and waste pollution from environmental public goods, the major resources of such pollution are analyzed; government policies and the main supply mechanism of environmental public goods are further researched to seek the root of problems, and to explore the factors constraining the supply of China's environmental public goods.
(3) Based on Diamond's OLG model, two supply modes of environmental public goods (i.e. supply by the government and government-led joint supply by enterprises and residents) are respectively introduced into the model to explore Pareto improvement, a comparative study is further carried out, and residents' marginal willingness to pay for environmental protection is measured. This is the core content of this book.
(4) According to different types of environmental public goods, an empirical analysis on the supply of a representative type thereof—clean air, is carried out to explore the methods of measuring residents' marginal willingness to pay for environmental protection, and the comparative advantages and disadvantages of three methods including revealed preference, stated preference and life satisfaction measurement, as well as their applicability in China are evaluated. This is the methodology and major empirical study of this book.
(5) According to the context of "Chinese-style Decentralization", the incentive mechanism for government provision of environmental public goods and the behavior characteristics thereof are explored, and a complete set of policy schemes for improving the production efficiency of environmental public goods and benefiting future generations are proposed.

The research framework hereof is depicted in Fig. 1.1.

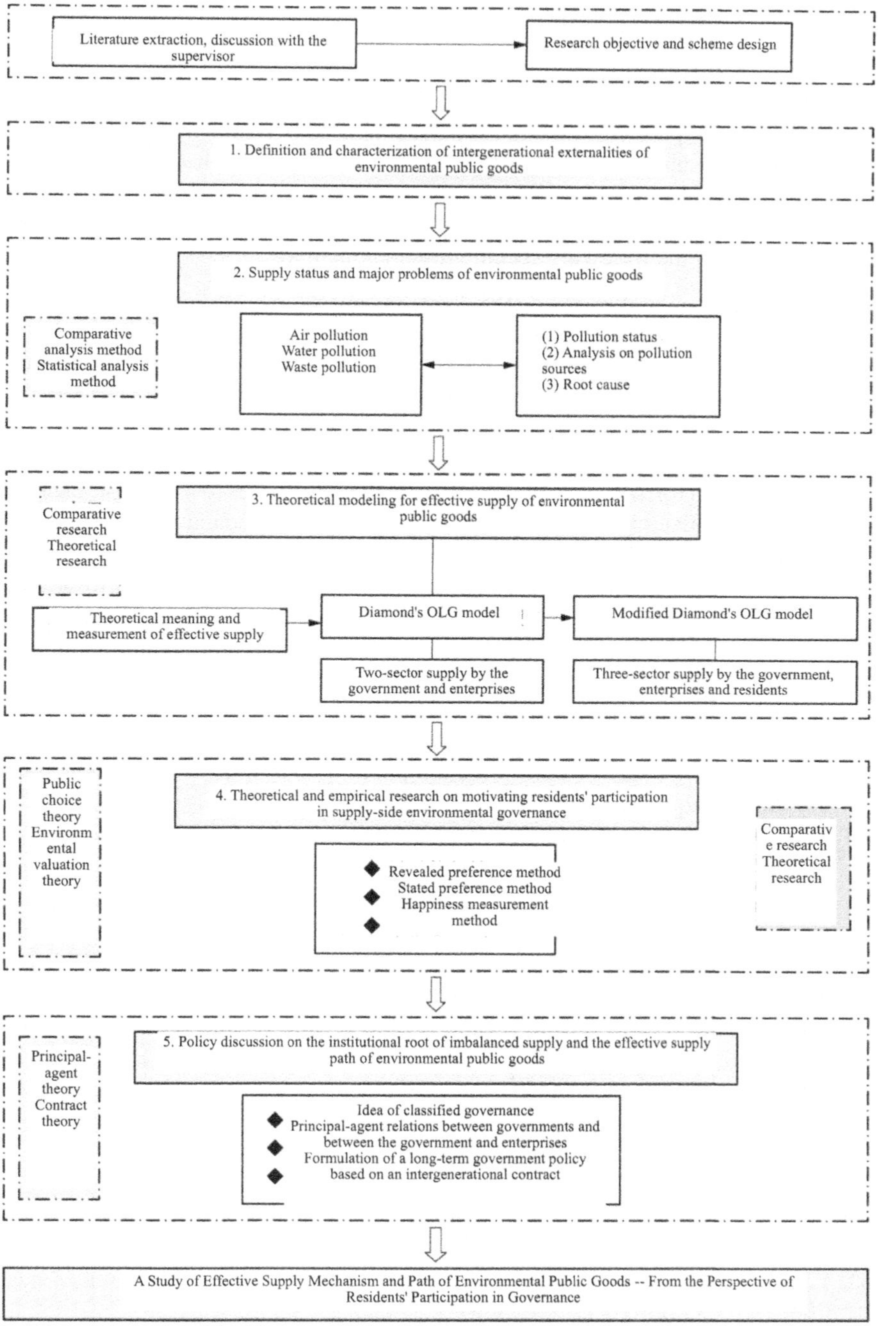

Fig. 1.1 Research framework

1.3 Research Methods

This book performs logical deduction and theoretical modeling on the basis of literature research, empirically analyzes the deduction and argumentation results, and further improves and refines the model. Then, it systematizes theoretical methods, and normalizes policies and measures to form a design idea for the effective supply path of environmental public goods conforming to China's national conditions. We have summarized the following specific research methods.

(1) Theoretical research method. This book carries out a theoretical research on the basis of literature review and research. Based on Diamond's OLG model, in accordance with the idea of the intergenerational externalities of environmental public goods, it probes into two supply modes of intergenerational public goods, i.e. supply by the government and government-led market supply, explores the feasibility of Pareto improvement; further researches the happiness measurement method for valuation of environmental public goods to provide a scientific basis for motivating residents to participate in environmental governance.
(2) Empirical research method. This book employs econometrics and public economics to empirically research residents' marginal willingness to pay for three representative types of environmental public goods, respectively air, water and waste pollution, and probes into feasible plans for applying theory into practice.
(3) Literature research method. Consulting relevant literature, we analyze the existing researches on the supply modes of public goods, and reveal the problems in the existing mode of environmental public goods so as to lay a foundation for the subsequent specific analysis.
(4) Comparative research and empirical methods. On the basis of reading up on literature and relevant research reports, we compare the differences in marginal willingness to pay between developed and developing countries, further explore the institutional root causing the misallocation of China's environmental public goods, identify the mechanism where Chinese residents' marginal willingness to pay is lower than that of developed countries; and comparatively analyze the supply paths of environmental public goods in different regions to determine a preferable scheme.

1.4 Main Content and Innovations of the Study

1.4.1 Main Content of the Study

1. Analysis on the double (intergenerational and spatial) externalities and supply status of environmental public goods

Clarifying the double characteristics of spatial and intergenerational externalities of environmental public goods is conducive to the subsequent further research on the supply mechanism thereof. The spatial externality means that an economic subject produces an external influence on another one, but the external influence cannot be traded at market price. The intergenerational externality means that the investor and the enjoyer (beneficiary or sufferer) of public goods are not in the same period of time, i.e. the effect duration of intergenerational public goods exceeds the investor's lifespan, and such goods belong to special public goods shared by more than one generation. The non-excludability and non-rivalry of intergenerational public goods not only exists in space, but also stands out in time. The characteristic of public goods generating spillovers in different periods of time is known as the intergenerational externality thereof [18]. Environmental public goods have typical intergenerational externalities, thereby increasing difficulty in the supply of such public goods.

This book systematically analyzes the supply status of China's environmental public goods, and probes into the supply mechanism thereof conforming to China's practical conditions. Focusing on air, water and waste pollution, we have carried out horizontally and vertically comparative researches on China and other countries, found that environmental public goods are in short supply under the current supply mechanism, and further explored the factors constraining the supply thereof in China.

2. Building of the OLG model for residents' participation in the effective supply of environmental public goods

In light of different supply mechanisms and modes of public goods concerning government supply, market supply and voluntary supply, based on Diamond's OLG model, this book first builds the two-sector OLG model that only covers the government and enterprises, then introduces residents into the model to theoretically deduce the modified three-sector model, and the theoretical mechanism that motivates residents to shift from the demand side of quality environment to the supply side so as to promote the transformation from a "visible hand" to an "invisible hand", to achieve an equilibrium from parts to a whole, and to finally realize Pareto improvement and the elevation of social welfare.

(1) Two-sector (government and enterprises) supply model (Diamond's OLG model)

Environmental protection has typical characteristics of public goods. Its non-rivalry on consumption and non-excludability on benefits make it impossible to truly embody effective demands, and difficult to avoid the "free-rider" problem in the supply process. Air pollution has the typical characteristics of forward intergenerational public goods, with obvious generational and intergenerational externalities. Given that residents are influenced by the thought of "free-riding", we could only rely on the government to improve environmental quality by imposing pollution taxes on enterprises so as to rectify the external diseconomy on the market.

(2) Three-sector (government, enterprises and residents) supply model (modified Diamond's OLG model)

In the new historical environment where China's economic development mode is undergoing transformation, and the financial budgets of local governments at all levels are tending towards tightening, marketization is capable of bringing Pareto optimality for the supply of intergenerational public goods under strict conditions. However, Rangel's [19] research indicates that the provision of backward intergenerational public goods (e.g. social security) can effectively increase the supply of forward intergenerational public goods (e.g. environmental protection). On this basis, in the context of the contract theory, the principal-agent theory, the public choice theory, and other theories describing China's reality, this part deepens the above two-sector supply model, and attempts to build a model that is more consistent with China's national conditions, and to draw a meaningful conclusion. To reduce residents' "free-riding" behavior and facilitate the sustainability of consumption, the government encourages residents to actively participate in environmental governance, and still imposes a carbon tax on polluters (pollutant discharge enterprises) for the purpose of remedying market failures.

3. Feasibility research on residents' participation in the supply of environmental public goods

How to motivate residents to participate in the governance of environmental public goods is an important consideration for system design and policy selection herein. This book first compares three environmental quality assessment methods (revealed preference, stated preference and happiness measurement), then describes the advantages of happiness measurement in measuring residents' willingness to pay for environmental protection, and further analyzes the feasibility of residents' participation in environmental governance by a comparison of material life conditions, technological level and residents' environmental awareness.

(1) Traditional methods of measuring environmental quality assessment

The traditional methods of environmental quality assessment comprise the revealed preference method and the stated preference method [20]. The former includes the protective expenditure method, the travel cost method, and the hedonic pricing method, which deduce the people's valuation on environmental quality based on their market conduct and payout. However, the protective expenditure method only considers the direct cost incurred from environmental pollution, neglects its indirect cost (e.g. damages to landscape), and thus undervalues environmental quality; the travel cost method is only applicable to tourist attractions, and has a high implementation cost; the hedonic pricing method relies on the existence of Walrasian general equilibrium, but ignores asymmetric information, migration cost and other market imperfection factors. The latter mainly includes the contingent valuation method, which requires respondents to directly give economic assessments by positing an environmental change. This method is widely applicable, but likely to result in biased or inaccurate estimations by placing the respondents in an imaginary alien environment.

(2) Measurement of residents' willingness to pay for environmental protection—happiness measurement method

The happiness measurement method, also known as life satisfaction measurement method, i.e. deducing the substitution relations between environmental quality and income level based on the happiness decision function, was proposed by German scholar Welsch [21] earlier. Compared with traditional methods of environmental quality assessment, happiness measurement has prominent characteristics. It deduces the "price" of individuals' willingness to pay by directly observing the utility level (subjective happiness), overcomes the theoretical obstacle that public goods have no market prices because of no market transactions to some extent, and thus appears more direct than the revealed preference method; meanwhile, this method does not require respondents to directly undertake environmental quality assessment, and only needs a little information concerning respondents' subjective happiness and local environmental quality, so it is more indirect than the stated preference method. In short, the happiness measurement method is an important supplement to traditional methods [20, 22].

(3) Resident's material life conditions and environmental awareness

This book further describes the feasibility of resident's participation in environmental governance by a comparative research on Chinese rural and urban residents' real income, Engel's coefficients and China's per capita GDP; uses the statistical data from CGSS to analyze Chinese residents' environmental awareness and knowledge, and to smooth the way for motivating residents to participate in environmental governance subsequently.

4. Empirical research on motivating residents to work on the supply side of environmental governance

The foregoing paragraphs have indicated that residents' participation in environmental governance can elevate social welfare through theoretical modeling. Taking air, water and waste pollution control as an example, this part employs the happiness measurement method to introduce the model of government-led pollution control where residents and enterprises jointly participate (three-sector model), and empirically analyzes that residents can shift from the demand side of clean environment to the supply side during pollution control on certain conditions.

In the empirical analysis, the real estate issue, as a concern of residents, is introduced into air pollution control to research residents' willingness to pay for environmental protection, mainly for the purpose of demonstrating that it is feasible to motivate residents to work with enterprises on government-led pollution control regarding environmental public goods. Besides, the specific currency amount paid for environmental protection can be measured by the happiness measurement method so as to enable heterogeneous individuals to make differentiated payments for environmental protection.

Through a study on the case of "Five Water Co-governance" in Zhejiang, this book indicates that it is possible for residents to participate in the supply of environmental public goods under the guidance of the government, and they may have a high enthusiasm for participation, provided that there are proper measures, as well as clear rewards and punishments.

Then, the significance of the supply mechanism is further demonstrated by the example of waste sorting in Shanghai. The government's effective policy is an important part of motivating residents to participate in the provision of environmental public goods.

5. As regards the policy research on the institutional root of misallocation of environmental public goods and the effective supply mechanism thereof, the contract theory is used. Focusing on the intergenerational externality of environmental public goods, this book designs an incentive reward and punishment mechanism applicable to intergeneration, and associates the present generation with future generations. To reduce residents' "free-riding" behavior, the government encourages residents to actively participate in environmental governance; still imposes a carbon tax on polluters (pollutant discharge enterprises) for the purpose of remedying market failures, applies the environmental tax levied to improving environmental quality, and gives incentives to enterprises which perform well in pollution control by means of offering red envelopes (financial incentives), e.g. offering loan support, with the amount equivalent to that of tax.

(1) Institutional root of misallocation of environmental public goods and theoretical mechanism thereof

Starting with the Chinese decentralization system characterized by "Political Centralization and Economic Decentralization", this part researches the root causing the distortion of local governmental actions and the misallocation of IPGs, and explores the specific theoretical mechanism therein according to the following logic.

First, characterization of "Chinese-style Decentralization". The core characteristic of "Chinese-style decentralization" lies in the close integration of "Economic Decentralization" and "Political Centralization". "Economic Decentralization" means that local governments are responsible for developing local economic policies and providing local public goods (more than 80% of the nation's financial expenditure is local government expenditure); "Political Centralization" means that the central government maintains politically vertical management, and has the power to appoint and remove local government officials to a great extent. The characterization of "Chinese-style Decentralization" is the premise of policymaking.

Second, principal-agent relations between governments. This part focuses on clarifying the financial power relations between the central government and local governments.

Third, supply behavior and allocation efficiency of environmental public goods in the decentralization system. This book focuses on analyzing how to improve the fundamental institutional arrangement in the new historical context where China's economic development mode is undergoing transformation.

(2) Policies for effective supply of environmental public goods

Through the above research, we identify the root of misallocation of environmental public goods. Similar to Tiebout's mechanism, local governments may compete fiercely for performance, but the difference from Tiebout's mechanism lies in the fact that due to the restriction of China's household registration system on the free migration of population, motivated by the behavior of "being responsible to superiors", local governments are more inclined to stimulate the provision of intergenerational public goods, e.g. construction of infrastructure like roads and bridges, which are capable of promoting economic growth in a short term (or within a term of office), and relatively neglect those capable of improving the people's livelihood, e.g. environmental protection, education, medical and health care, and other public services, which are the essential conditions of boosting the long-term sustainable economic growth. In the face of economic growth accompanied by high environmental costs, how does the government effectively intervene in governance? How should a policy for market-oriented supply of environmental public goods be developed? Which supply mechanism and path should be chosen?

1.4.2 Characteristics and Innovations of the Book

1. Environmental public goods are typical intergenerational public goods. The imbalanced supply of intergenerational public goods leads to the serious under-supply of forward intergenerational public goods. Starting from the rational man hypothesis, the present generation makes decisions on their own side, with little consideration for the interests of future generations. As a result, compared to the backward intergenerational public goods which provide old-age security for the present generation, the forward intergenerational public goods to the benefit of future generations are in seriously short supply. This study provides feasible plans for the effective supply of forward intergenerational public goods.
2. The effective supply model of environmental public goods is applicable to the analysis on the effective supply path of China's intergenerational public goods. Diamond's OLG model is used to research the overlapping generations economy. In consideration of dynamic infinite overlaps, this study makes improvements on

the basis of the original model, probes into two supply modes of environmental public goods, i.e. supply by the government, and government-led supply by enterprises and residents, explores and compares them to determine which one is capable of realizing Pareto improvement. This is a theoretical innovation.

3. This study employs the happiness measurement method to theoretically probe into the feasibility of residents' participation in environmental governance, and empirically analyzes Chinese residents' marginal willingness to pay for environmental protection to explore the institutional root of misallocation of environmental public goods and the theoretical mechanism thereof, thereby providing a scientific decision-making basis for the supply of environmental public goods in developing countries.
4. The institutional arrangement of "Chinese-style Decentralization" is a significant root of misallocation of China's intergenerational public goods. The decentralization system with the core characteristic of "Political Centralization and Economic Decentralization" causes local governments to devote themselves to observable and comparable tasks including economic growth, but ignore education, environmental protection, healthcare and other livelihood improvement work, which can be hardly observed and compared. Starting from the intergenerational externality of environmental public goods, this book motivates residents to participate in environmental governance, encourages the government to introduce sustainable development policies, researches the "pollution co-governance" by the government, enterprises and residents, and further uses the contract theory to put forward policy suggestions. This is an innovation on research strategies.

Chapter 2
Theoretical Basis and Literature Research

2.1 Theoretical Basis of the Study

The study of environmental public goods is connected with many theories concerning intergenerational externality, altruism; public goods supply mechanism, public choice, contract and motivation.

2.1.1 Public Goods and Intergenerational Public Goods Theories

The theoretical research of public goods derives from the definition of a classic theory presented by Samuelson [23], who classified social goods into public and private goods. In accordance with the definition given by Samuelson in *The Pure Theory of Public Expenditure*, a pure public good or service means that the consumption of such good or service by one person does not decrease the consumption by another person. The two characteristics—non-rivalry and non-excludability are mainly analyzed. On the one hand, non-rivalry means that marginal costs are zero, i.e. the addition of one consumer does not increase the supplier's marginal costs; on the other hand, it implies that marginal congestion costs are zero, i.e. the consumption by one consumer does not affect the quantity and quality available to other consumers.

Among public goods, some goods or services have an intergenerational transition function. With such a characteristic, environmental public goods are a type of intergenerational public goods. Xiong [24] held that the non-excludability and non-rivalry of public goods not only existed in space, but also stood out in time. She described the spillover characteristic of public goods in different time intervals as the intergenerational externality thereof.

J. Yang, *Effective Supply Mechanism and Path of Environmental Public Goods in China*,
Public Economy and Urban Governance in China,
https://doi.org/10.1007/978-981-99-2939-9_2

After the development of the concept of "Intergenerational Goods", Cornes and Sandler [25] researched intergenerational externality and presented the "Intergenerational Club" theory, believing that club members had to bear an important cost—temporal "depreciation arising out of use". Based on this characteristic, they applied an "inter-temporal" (intergenerational) model to test the intergenerational characteristics of public goods, discussed the factors influencing the supply of intergenerational public goods, and especially analyzed the effect of government behavior on the supply thereof. The "depreciation arising out of use" occurs between present and future generations, so the intergenerational model is based on the important precondition of acknowledging the existence and continuity of future members. In light of the intergenerational club theory, the "inter-temporal" (intergenerational) model is built, and depreciation function is introduced to reflect the temporal characteristic. On the contrary, traditional models do not take temporality into account in the process of analyzing problems relating to public goods. This is an outstanding contribution of this theory.

Lowry [26] held that intergenerational public goods were goods produced or services provided for future generations, e.g. a portion of land retained and preserved for future generations on certain natural conditions, social security, retirement pension, tax incentives set for those who recycled or developed renewable energy sources, and long-standing space exploration. Zhao [9] pointed out that intergenerational public goods were public goods objectively existing in the context of overlapping generations economy, i.e. goods shared and used by more than one generation. Compared to generational public goods, they have an obvious intergenerational externality, which is specifically manifested in that the supply and consumption subjects are beyond the scope of one generation, and significant non-excludability and non-rivalry exist between generations. Intergenerational public goods refer to public goods in connection with the investment-return relations between two or more generations, serving as an extension of public goods on the temporal dimension in a general sense.

Due to the abstractness and complexity of the concept of intergenerational public goods, there are lots of literature documents expounding thereon, with the listing of different classifications. The classification by "generation" is relatively accessible, which means that public goods having an effect on more than one generation are intergenerational public goods. Here the main classification standard is the determination of a "generation", including defining a decade as a generation in the demographic sense; defining a governmental term of office (5–6 years) as a generation; defining old, middle-aged and young people as three generations or old and young as two generations in the family mode.

According to Buchanan's "Club Goods" theory and Olson's "By-product" theory, any public goods are associated with an organization, and belong to the members therein. After the "Intergenerational Club" theory was presented by Sandler, numerous scholars introduced the aforesaid "Organizational Series" thoughts into the classification of intergenerational public goods, which could be classified by the "organization" associated therewith. One level of a complete organizational series consisting of various social organizations is the smallest public organization (composed of two persons); the other level is the largest organization (supranational global organization). All sorts of social organizations are arranged from small to large between the two levels. For example, the boundary of family education is family, which is focused on the education investment and enjoyment between generations of a family; environmental pollution and similar intergenerational public goods are cross-border, which are focused on the provision of such cross-border intergenerational public goods among different countries.

One of the common classifications is based on the utility direction of intergenerational public goods [19]. In case that we invest in some public goods which have an effect on the future generation (children) or parents, with the present generation as the object of study, the public goods effective for the previous generation after investment are known as backward intergenerational goods (BIGs), e.g. social security; those effective for the future generation are known as forward intergenerational goods (FIGs), e.g. public education and environmental protection. With the help of the concepts of positive and negative externalities, we may also classify intergenerational public goods into good and bad ones, generally referred to as intergenerational public goods and bads. For instance, environmental protection and R & D belong to intergenerational public goods; bad habits and political systems belong to intergenerational public bads.

Another classification is based on the cause of intergenerational externality of public goods—the "generation" to which the investor and the beneficiary belong. When the investor and the beneficiary belong to different "generations", such public goods as education and old-age security may bring benefits to the future or previous generations upon investment by the investor, and the supply thereof may rely on a credit mechanism. In this case, such public goods are known as credit-type intergenerational public goods. As regards the other type, in the event that the investor and the beneficiary belong to the same "generation", and the effect duration of public goods exceeds one generation's (investor) lifespan, the future generation (beneficiary) will enjoy the benefits by inheriting public capital stock. We describe such public goods as durable intergenerational public goods, e.g. highways and public buildings [27].

We may classify public goods into institutional and material intergenerational public goods by the attributes thereof. For example, national political systems belong to the former; environmental protection, education, and the like, belong to the latter.

The review of types of intergenerational public goods is convenient for an effective study. It is easy to see that as the investor and the beneficiary of intergenerational public goods do not exist in the same time and space, intergenerational externalities will arise inevitably. Thus, the study of environmental public goods herein is mainly focused on the analysis of intergenerational externalities.

2.1.2 Public Goods Supply Mechanism Theory

The public goods theory is a collection of major theoretical results in public economics. The earliest theoretical result is the Lindahl equilibrium model of 1919. In the view of Lindahl, regarding the supply of public goods, individuals have the equal right to purchase the sum of public goods at a price they are willing to pay. Thus, on the condition of tax negotiation, the government could obtain a Lindahl tax system to ensure the supply of public goods.

As regards general public goods, Musgrave [28] and Samuelson [29] both held that due to the non-excludability and non-rivalry of public goods in consumption, private supply would definitely have efficiency or benefit losses, thus, they must be supplied by the government. Pigou modified the external effect of common resource consumption by means of Pigovian taxes and other measures to effectively coordinate the interest relations between different subjects in economic activities, and stressed that the internalization of external effect was the most effective measure to solve externality. Subsequent literature researches demonstrate individuals' potential in provision of lighthouses, education, laws and orders, infrastructure, agricultural scientific research and other public goods. In terms of the supply paths of intergenerational public goods, government supply and private supply, which one is more effective? Or is the joint supply by the government and individuals more effective?

The supply mechanism of public goods is a supply mode abstracted from the perspective of the supply subject and the operation mechanism. Influenced by neoclassicism in the early period, Western scholars held pessimistic views on the supply and governance of public goods. Public goods have no clear property rights, leading to externalities and incapability of effectively preventing users from "free-riding", and thus are inevitably in short supply. Similarly, due to the unclear definition of property rights of public open resources, motivated by self-interest, individuals will definitely excessively rake in the benefits from common resources, thereby resulting in rent dissipation. The problem is always closely connected with the cooperation, "free-riding" and collective action in the social dilemmas between public and social choices. This is dubbed as "Tragedy of the Commons" by Hardin, and "Prisoners' Dilemma" in the game theory. Olson summarized the process of individual rationality leading to collective irrationality into the "Logic of Collective Action". Rational resource users are prone to the irrational collective action problem, rather than cooperation.[1] The study of this part includes four classic models: Olson's [16] "Logic of Collective Action", Dawes' [30, 31] "Prisoners' Dilemma", Hardin's [32] "Tragedy of the Commons", and Ostrom's [15] "Governing the Commons". Among them, Hardin's "Tragedy of the Commons" and Dawes' "Prisoners' Dilemma" both demonstrate that a decentralized individual decision-making system is incapable of effectively supplying public goods, and rational individuals have no possibility to cooperate. Economists often use them to prove the necessity of government supply and the failure of market supply. Government intervention is no more than direct

[1] Yan Jia, Zhang Ping, Xu Zhicheng. Research on Voluntary Provision of Public Goods in the Dual Incentives of Institution and Social Preference, Soft Science, 2014, Issue 2.

government control, and the common means thereof is command and control. For externalities, the government may also adopt means to give incentives to enterprises with positive externality; impose tax penalties on those with negative externality, and ensure the equivalence between marginal social benefits and costs by tax regulation. However in fact, the government is unable to master all of the relevant information to develop proper policies, and may be disturbed by related interest groups. Decades of development demonstrates that the actual effect of direct government intervention is unsatisfactory, and government intervention may result in various failures. Another mode is market supply, which is focused on privatizing the property rights of public goods. Coase, one of the representatives, proposed that in the case that property rights were clearly defined and available for free transaction, the final result would be Pareto optimal, no matter who owned the property. Therefore, as long as property rights are clearly defined, the supply of public goods can be completed by individuals. The theory seems perfect, but has a prominent problem, i.e. the excessively high cost of property rights transaction is likely to destroy the effectiveness of the governance mode.

After a detailed empirical analysis on the police service in metropolitan areas, Ostrom found that "there are not just two orders—government (Thomas Hobbes' theory of sovereignty) and market (Adam Smith's market theory) in public economics". Later, Olson first carried out a pioneering analysis on the theory of voluntary supply of public goods, answered the question as to why some organizations were capable of providing public goods in *The Logic of Collective Action*, and found that small groups were more likely to reach the supply goal. Bergstrom et al. [33] probed into the effect of different degrees of wealth redistribution on individual donations. Their analysis on a general model of non-cooperative voluntary contributions is a classic analysis on voluntary supply of public goods. In *Governing the Commons*, Ostrom appeared optimistic. She used lots of cases for empirical analysis, and researched how the principal enabled everyone to obtain lasting common interests by self-governance in the face of temptations concerning "free-riding", responsibility evasion or opportunistic behavior. Fan [34] anatomized the government supply, market supply and voluntary supply mechanisms of public goods, and considered that the market supply mechanism of public goods was a mechanism where profit organizations made up for expenditure by charges for profit purposes; the government supply mechanism belonged to a "second-level" mechanism, which was based on resource allocation in the market, aimed at fairness, and took taxes and public charges as main fund-raising means; the voluntary supply mechanism belonged to a "third-level" mechanism, which implemented resource allocation on the basis of the market and government mechanisms taking effect, and took donation as the main mode, with an altruistic aim.

2.1.3 Altruism Theory

Due to the "free-rider" problem of public goods, the theory of egoism is actually incapable of interpreting the private voluntary supply of public goods. In another masterpiece *The Theory of Moral Sentiments*, Adam Smith held that "compassion" stimulated altruistic behavior, and used the principle of compassion to interpret the source of the sense of justice and other moral sentiments of mankind, construing it as the foundation sustaining society. He considered altruism—"obtaining joy from other people's happiness" as part of human nature.

Becker [35] presented the theory of altruistic motivation, built an OLG model with altruistic factors, and considered that the wealth transfers including selfless gifting between family members could maximize the overall family utility.

With the help of the rational man hypothesis in traditional economics, Becker [36] brought the utility of future generations into that of the present generation for analysis, and successfully incorporated altruism into the research paradigm of mainstream economics. It should be particularly noted that the altruism introduced by Becker refers to a behavioral outcome, i.e. the individual behavioral outcome is beneficial to other people, but the motivation thereof actually lies in individual rationality (self-interest). In his view, "The reason why altruism is less common in the market than in families is that it is inefficient in market exchanges, but efficient in family life."

In the view of Collard [37], individual unselfish behavior will bring individual or group interests, which are higher than costs. Andreoni [38] presented "warm-glow", believing that economic man gained utility improvement from the donation behavior, and other people's donations could not entirely supersede his or her own donation behavior. The bounded rationality theory [39, 40] represented by Simon interprets altruism from the perspective of behavioral motivation. In his view, "If one sacrifices his or her own wealth or power for that of someone else, such behavior is altruistic; if one seeks to maximize his or her wealth and power, such behavior is selfish." In the eyes of Andreoni [41], when individuals are purely self-interested, they always choose "free-riding" in the face of public goods. However in reality, the individual supply of public goods is not zero, and individual altruism does exist. When individuals behave altruistically, their utility level is not only the function of the total consumption of private goods, but also the function of that of public goods. In other words, individuals care about not only their own consumption of private goods, but also other people's consumption of public goods, with no concern for whoever donates public goods.

Some scholars proposed the "warm-glow effect" of common resources supply, i.e. individuals helped others by providing public goods, and accordingly obtained psychological compensation, which included status, prestige, and social recognition. After reviewing existing theoretical and empirical results, Hu and Wu [42] found that individual altruism was "extended self-interest" in essence, which meant no longer one's own interest, but interest related to oneself, including caring for parental generations, people around and society.

According to existing researches, scholars have classified altruism into relative altruism, reciprocal altruism and pure altruism [43]. Hamilton [44] indicated that relative altruism meant that individual organisms would make sacrifices for their blood relatives, and explained that species providing relative altruism had a significant evolutionary advantage in the struggle for existence. Additionally, Hamilton proposed the theory of kin selection, i.e. the friendliness between different individuals depended on the proportion of "consanguinity" shared by them. As regards individual altruistic behavior, basically, natural selection acts on genes, which are not altruistic. Hamilton successfully interpreted the relative altruism ubiquitous in the biological world from an individual perspective.

In the view of Trivers [45], reciprocal altruism means that individuals, who are unrelated by blood, help each other in order to get return. Axelrod and Hamilton [46] developed the theory of evolution of cooperation, and interpreted the reciprocal altruistic behavior under a self-interest assumption by applying the game theory method. Rabin [47] employed the psychological game framework to build a motivation-based theoretical model of economics for reciprocal preference in a pioneering manner, defined reciprocity as a desire to "return good for good, and evil for evil", and specifically analyzed the widespread reciprocal behavior in human life.

Premised on complete information, the inequity aversion model presented by Fehr and Schmidt [48] assumes that the people may damage their utility level because of income inequality. In their model, participants care about not only their own interests, but also the income difference between them and other people. This theory assumes that the people have an aversion to income inequality to some extent, and their aversion against disadvantageous inequality is stronger than that against advantageous inequality. If the theory is applied to public goods, as long as inequality abhorrers believe that others will make voluntary contributions, they will do so as well. Bolton and Ockenfels [49] presented an ERC (equity, reciprocity and competition) model based on an incomplete information assumption. They believed that individual behavior was motivated by both absolute and relative returns, and constructed the utility function, known as motivation function.

Edwards [50] offered a biological interpretation of pure altruistic behavior, proposed the group selection theory, and held that genetic evolution was realized at the biotic population level; the behavior characteristic might be preserved and evolve with the maximization of population interests, when individual organisms' altruistic behavior was advantageous to such interests.

2.2 Research Review of Environmental Public Goods

Focusing on the research topic, this book mainly conducts a literature review on the characteristics of intergenerational externalities of environmental public goods and the supply modes thereof.

2.2.1 Analysis on Characteristics of Intergenerational Public Goods

In view of intergenerational characteristics, the following problems are mainly discussed in the academic circles, with an attempt to crack the internal characteristics of intergenerational public goods in a clearer way.

1. Intergenerational externalities

The intergenerational positive externality of environmental public goods is manifested in the fact that the present generation undertakes the investment cost of public goods, and the future generation can obtain additional revenue at no cost [66]. Conley mainly researched the intergenerational externalities of intergenerational public goods produced, used or transferred in the form of materials or assets, i.e. the "theory of intergenerational externalities of public goods". If intergenerational externalities cannot be internalized to a certain extent, such intergenerational public goods will face systematic undersupply. In his view, the provision of public goods by local governments is capable of fully realizing the internalization of intergenerational externalities. On the contrary, the central government is incapable of internalizing intergenerational externalities by providing public goods in accordance with the national average of demand. In addition, the existence of interregional externalities offsets the benefit effect of providing public goods by decentralization, and leads to efficiency losses. Considering the above factors, the government sector must design the supply mechanism of intergenerational public goods according to the relative intensity of spillovers between administrative regions and intergenerational externalities, and make a reasonable choice between the two institutional arrangements of central government supply and local government supply.

Li [66] proposed that the behavioral externality of government supply of public goods could be divided into two types: spatial externality and temporal externality. For instance, the positive externality of time is manifested in the fact that the present generation undertakes the investment cost of public goods, and the future generation can obtain additional revenue at no cost. Guo and Xia [67] pointed out that the main reason for the existence of intergenerational externalities lies in the incompleteness between generations. The "Transaction Insider" and the "Transaction Outsider" between generations have no direct transaction agreement arrangements to coordinate their rights and interests, resulting in the externality problem. And the primary cause thereof is the incompleteness of property rights between generations. Li and Sun [68] proposed that as for intergenerational public goods, the positive and negative effects of intergenerational externalities on the public in the same space varied with time, which were difficult to define. As regards the intergenerational public goods producing a positive externality on the present generation of residents, their intergenerational externality on the future generation may be either positive or negative.

2. Absence of the agent of future generations

Rangel [19] mentioned that when the behavior of the present generation occurred, the members of future generations had not yet been born, and thus could not negotiate or communicate with the present generation; accordingly, the present generation and the next generation(s) could not reach a binding contract to realize the supply of public goods. As future generations had no right to vote or negotiate, both representative democracy and direct democracy might have shortsightedness arising from the absence of the agent of future generations in terms of the supply of intergenerational public goods in the public choice process.

Related rights and interests arising from the behavior of the previous generation, which originally belonged thereto, might occur among future generations. Because of the intergenerational barrier on time, the previous generation cannot own such rights and interests. It is manifested as the absence of future property rights of the present subject. Due to the absence of property rights between generations, the behavior of the previous generation as the "Transaction Insider" produces a factual impact on the rights and interests of future generations as the "Transaction Outsider" [68]. Guo [69] indicated that if intergenerational rights and interests subjects reached a direct transaction agreement by other means, rather than by the property rights system, the cost thereof would be infinitely huge. Because of the absence of the rights and interests subject of future generations, they are incapable of reaching a direct transaction agreement with the present rights and interests subject. Accordingly, the cost of a direct transaction agreement between present and future rights and interests subjects would be infinitely huge.

3. Depreciation effect of environmental public goods

The inter-temporal depreciation characteristic of environmental public goods is referred to as depreciation effect. Following Sandler's theory of intergenerational clubs that takes into account the inter-temporal depreciation characteristic of intergenerational public goods, after researching material intergenerational public goods, Li Yufang pointed out that when the effect duration of public goods invested in by the present (or previous) generation exceeded the lifespan of one generation, future generations would enjoy the benefits by inheriting public capital stock. Such public goods are characterized by durability and resistance to consumption, and their own value and use value will be consumed over time. He proposed that the "depreciation rate" and the "retention rate" must be considered in the research on such public goods. In the view of Li and Sun [27], such phenomena as consumption impairment and devaluation arising from depreciation may occur in the forms of the loss of operating efficiency, the decline of reproduction capacity, or the decrease of available areas.

4. Intergenerational equity

Intergenerational equity means that the intergenerational allocation of resources maintains a fair relation, and forms a fair, reasonable consumption relation between generations, with no occurrence of too much inefficient consumption or too little insufficient consumption in a period. As the founder of the theory of intergenerational equity, Weiss [70] presented the trustee-principal relation between present and future generations. The present generation, on the one hand, has the right to use the resources (e.g. culture and environment) left by predecessors; on the other hand, has a duty to look after and maintain these resources so as to fulfill the trustee's responsibility. American scholar Page [71] advocated intergenerational equity. In his view, intergenerational equity can be summarized as follows: "Assuming that the outcome of the present decision will influence the interests of several generations, fair distribution should be implemented among these generations based on the above outcome. "Pearce [72] indicated, "According to a broad definition, sustainable development includes the provision of a certain quantity of quality wealth inheritances to the next generation, which should be at least equal to those inherited by the present generation. "Weiss [73] reanalyzed the relations between intergenerational equity and public policies (including environmental policies).

2.2.2 Research Path of Environmental Public Goods

For the research on the supply of intergenerational public goods, most of literature documents start from the hypothesis of economic man, with a focus on the analysis of intergenerational externalities, i.e. the externalities of the present generation extending to the later or future generations, which include intergenerational external economy and diseconomy. Scholars have built a large number of theoretical models, focusing on the following topic: the intergenerational externalities of intergenerational public goods determine the dilemma of undersupply. If the present generation is "self-interested", why could they realize intergenerational transfer of public goods? Furthermore, which system can achieve the optimization of supply of intergenerational public goods? Taking this at the core, scholars probed into the characteristics of intergenerational public goods and the problems arising therefrom. Their research mainly includes the following aspects.

1. Optimal allocation of intergenerational public goods

The optimal allocation of intergenerational public goods is an extension of the traditional Samuelson [29] condition, using the Pareto optimality condition—the marginal rate of substitution between intergenerational public goods and private goods equal to the marginal rate of transformation, i.e. $MRS_{GP} = MRT_{GP}$, to probe into the supply optimization of intergenerational public goods. Wei and Ren [74] carried out an analysis from the perspective of utility function, and held that the cost-revenue functions of the "Transaction Insider" and the "Transaction Outsider" both changed due to the existence of intergenerational externalities, which changed the equilibrium point of resource consumption with regard to the revenue maximization of the two in the state of perfect market, thereby causing the movement of the equilibrium point for maximization of marginal revenue. In this case, Pareto optimality cannot be achieved, and the overall social efficiency declines accordingly. Doeleman and Sandler [75], as well as Kotlikoff and Rosenthal [76] researched the investment problem of intergenerational public goods in a finite OLG model. In their view, each generation's self-interest motivation tends to give rise to the under-investment in backward intergenerational public goods.

2. Sustainable development

Sustainable development includes the provision of a certain quantity of quality wealth inheritances to the next generation, which should be at least equal to those inherited by the present generation. The intergenerational externality is the root of unsustainable development, and one of the core problems regarding the realization of sustainable development is how to eliminate intergenerational externalities to the greatest extent. In case of intergenerational externalities, the previous generation as the "Transaction Insider" obtains additional revenue, with no need to bear the corresponding cost, which is borne by the future generation as the "Transaction Outsider". As a result, intergenerational resource allocation deviates from the requirements for sustainable development, leading to the unsustainability of intergenerational resource consumption, imperiling the sustainable development of society.

In Table 2.1, we sort out the research paths of intergenerational public goods, and carry out an analysis in the respects of research topics, research basis, research objectives and representative figures.

Table 2.1 Research paths of intergenerational public goods in economics [18]

Research topic	Optimal supply of intergenerational public goods	Family supply of intergenerational public goods	Sustainable development
Research basis	Egoistic motivation	Altruism theory	Intergenerational equity
Research objective	$MRS_{GP} = MRT_{GP}$ Pareto optimality (or improvement)	Maximization of the overall utility of different generations of "interdependent" persons	Fair distribution of intergenerational public goods to each generation
Representative figure	Samuelson	Becker	Weiss

2.2.3 Strategies for Supply of Environmental Public Goods

Due to the double problems—patial and intergenerational externalities of environmental public goods, the effective supply thereof appears doubly difficult. At present, the research thereon has three directions: "Pigovian Tax", supporting policies for positive and negative externalities, as well as intergenerational public goods.

The first one is the pure "Pigovian Tax". Due to market failures, scholars believed that the government provision of a quality environment is relatively efficient. Based on Pigou's [99] research, a large number of scholars tended to support the imposition of an environmental tax [100–102], and proposed that the carbon tax had double dividends, which could not only curb those economic activities harmful to the environment, but also further lower the efficiency loss of the tax system, and increase employment by cutting down the tax with a distortion effect, thereby indirectly leading to the increase of social welfare. Bovenberg et al. [103–107] researched the optimal tax rate setting of an environmental tax. Li et al. [108,109] discussed the environmental tax rate setting and supporting policies of China. Zhang and Jin [110] proposed that the pollution tax would be conducive to reducing spending on pollution and health, but a higher pollution tax could not be always better. In case the pollution tax rate is below 40%, it is conducive to increasing the consumption level; in case the rate is below 60%, it is conducive to increasing capital accumulation for healthy manpower, promoting economic growth and improving the benefit level. Xiao and Liao [111] proposed that the government investment in pollution control had a threshold condition, and there existed an optimal ratio between the investment in pollution control and the total tax revenue. Researching the degree of correlation between air/water pollution/infant mortality and environmental regulation policies, Greenstone and Hanna [112] concluded that India's environmental regulation policies had a significant effect on improving air quality, brought a significant decline in PM value, SO_2 content and NO_2 content, but an insignificant decline in infant mortality; however, the water pollution regulation policies had no observable effect. These views greatly challenge the past popular view that environmental quality depended on income.

The second one is supporting policies. Relevant researches are focused on seeking effective environmental policy instruments, and advocate developing supporting policies which encourage positive externality and restrain negative externality simultaneously. A number of scholars comparatively researched different policy instruments. Requate and Unold [113] concluded that tax policies were more effective than free and auctioned permits in terms of giving incentives to enterprises. Cherry et al. [114] comparatively researched the public acceptability of three environmental policies: environmental taxes, subsidies and pollution discharge quantity regulation. The results showed that environmental subsidies were significantly supported while quantity regulation received a larger number of objectors. The research conducted by He and Li [115] suggests that simply implementing energy taxes and environmental taxes for reducing pollutant discharge will produce a negative effect on macro economy, but the combined use of such taxes and other environmental policies is capable of obtaining double dividends. Designing a series of experimental schemes, Liu [116] held that the contradiction between individual rationality and collective rationality could give rise to the market defect in provision of public goods, and the results of his experiments suggested that the rational system design had the possibility to boost the successful provision of public goods. Yingna et al. [117] considered that it was unsuitable to bring the energy tax into the alternatives of China's environmental policies at the present stage. Dongdong and Yang [118] concluded that the energy tax policy lowered the economic growth rate and the social welfare level while reducing pollution. According to their research, in case of the double externalities—subsidies for emission reduction R & D and environmental protection in the economy, the government could achieve triple-win results in environmental improvement, economic growth and social welfare elevation by imposing a pollution tax and offering appropriate subsidies to promote the emission reduction R & D activities of enterprises. In the view of Zhang and Lin [119], the externally imposed incentive mechanism is capable of effectively curbing the people's free-riding, and improving the degree of cooperation and overall benefits.

From the perspective of intergenerational externalities, intergenerational contracts are designed in numerous researches to spur the present generation to give consideration to the health of future generations, and to carry out effective environmental protection. Dahlman et al. [60] introduced a sustainable "intergenerational contract" to solve the intergenerational problem, which means that the future generation may transfer part of their wealth in the form of money or goods to the previous generation, as financial compensation for the previous generation's performance of the contract. Fudenberg et al. [120] stated that the presence of sustainable intergenerational goods could be used to solve the moral hazard, and settle the inefficiency problem in the intergenerational economy. John [13] proposed that environmental taxes could facilitate the internalization of externalities in an OLG model containing environmental externalities. Environmental taxes enable consumers to reduce the provision of private goods with negative externality so as to maintain the amount of environmental pollution, and achieve the steady economic and environmental growth. Ono [121] designed optimal tax schemes (consumption tax and intergenerational income transfer) in an OLG model to internalize environmental externalities. Beginning with intergenerational externalities, Bovenberg et al. [105] conducted an analysis, concluded that the improvement of environmental quality was at the cost of the present generation's welfare losses while the future generation as the beneficiary did not bear the cost, and suggested that the government should develop a bond portfolio policy of deficit financing to achieve intergenerational transfer. Based on a game theory analysis, Rangel [19] considered that the agents of all generations provided the same level of goods to be transferred to the previous and next generations by the present generation in accordance with the goods transferred backward by the previous generation. As long as these intergenerational transfers grow in step with the interest rate, participants will all receive services at an optimal level and will not betray the strategy. Later, Rangel [122] researched how to protect future generations and provide optimal intergenerational public goods by introducing tax-base restrictions to directly translate intergenerational externalities into one of the variables influencing present generations, in an attempt to motivate them to show concern for future generations. Based on an OLG model, Ono [123] researched the effect of environmental taxes on welfare distribution in the presence of involuntary unemployment, and believed that environmental taxes could improve environmental quality, employment and social welfare within a generation, but might yield a trade-off across generations; the improvement of the present generation's welfare level had to be at the cost of the decline of the future generation's welfare. Based on Diamond's OLG model, Li [68] analyzed the competitive behavior among local governments, which could capitalize and internalize intergenerational externalities through the price change of land within the area under administration, thereby effectively coordinating intergenerational benefit distribution and improving the supply level of intergenerational public goods. Xiaoli [24] employed the game theory method to explore the game between intergenerational equity and environmental resources. In case the share of resources exceeds the threshold value, the welfare revenue from protection of environmental resources will offset the welfare loss arising from taxes. Thus, all generations prefer intergenerational transfer policies, rather than laissez-faire.

2.3 Several Comments on Research Literature

There are few researches on intergenerational public goods at home and abroad. Although the limited quantity of materials searched could give us some new ideas, they have the following deficiencies regarding environmental public goods.

First, particular emphasis is put on generational research, with a lack of intergenerational research. Most of the existing literature is limited to the generational discussion on spatial externality. However, environmental protection belongs to typical forward intergenerational public goods with a significant effect on the next generation and even future generations, so it should be treated and analyzed in the long run. How do we conduct a theoretical review and make a breakthrough in the absence of agents of future generations and the presence of intergenerational externalities? This is an immediate problem for the study of pollution control.

Second, particular emphasis is placed on partial research, with a lack of overall research, and little research on motivating residents' participation. Among the intergenerational discussions on environmental public goods, the discussions based on an OLG model are basically focused on the analysis of governments and polluters, but neglect the fact that residents as the demand side of clean environment are actually a source of pollution, e.g. vehicle emission exhaust, household waste pollution and water pollution. If all the subjects of society (the government, enterprises and residents) are brought into the supply side of clean environment, can the modified OLG model reach equilibrium? Does it reach the Pareto improvement compared to the equilibrium only with the participation of the government and enterprises? This requires extending partial research to overall research for analysis.

Additionally, the research with residents as the supply side of clean environment is conducive to not only effectively supervising government and enterprise behaviors, but also motivating residents to restrain their behavior: water conservation, green travel, and environment-friendly lifestyle. However, such research is rare at present.

Third, the happiness measurement method for valuating environmental public goods lacks the empirical validation from developing countries. Most scholars analyzed Western countries, but seldom probed into the supply of intergenerational public goods in the context of China. So far, most of the researches on environmental public goods are focused on measuring residents' willingness to pay for environmental quality, and the method thereof gradually shifts to happiness measurement. This provides a scientific basis for the government to motivate residents to actively participate in environmental governance, but the academic circles have no in-depth exploration with regard to the factors influencing residents' willingness to pay, with few researches focused on China [124].

Fourth, the institutional root of misallocation of environmental public goods and the theoretical mechanism thereof fail to be explored thoroughly. The existing literature basically reaches a unanimous conclusion that China's market-oriented reform and opening-up significantly improves the allocation efficiency of private goods, but China has the phenomena of long-term undersupply and misallocation in the public sphere, especially in terms of intergenerational public goods with obvious temporal characteristics, e.g. environmental protection, basic education, medical and health care. There is little literature further exploring the institutional root of misallocation of environmental public goods. Moreover, some researches find that Chinese residents' marginal willingness to pay is lower than that of developed countries, but fail to identify the specific mechanism thereof, or present any effective solutions.

All these deficiencies can hardly solve the current problem of environmental quality improvement. No matter from the generational or intergenerational perspective, the existing research of environmental public goods is focused on government policies. However, in the pollution control process, the government lays emphasis on polluters, and controls pollution mainly by means of imposing environmental taxes. Even in terms of multipronged policies, taxes and subsidies are targeted at polluters. However, as the demand side of clean environment, residents are keenly aware of the environmental impact on health, concerned about the quality of future generations' living environment, and subjectively willing to assume some social responsibilities.[2] How do we enable residents to become the supplier of quality environment from the perspective of intergenerational externalities? How does the government design an appropriate policy to motivate all people to participate in environmental governance? This is not only an area that has not been researched by predecessors, but also a new attempt for the "supply side" reform advocated at present. This book attempts to make some breakthroughs in these aspects. Based on Diamond's OLG model, starting from the intergenerational externalities of environmental public goods, this book extends the environmental quality research models introduced by John and

---

[2] According to the CGSS of 2010, as regards the answers to the question: "To what extent are you willing to pay higher prices for environmental protection," 60.65% of the respondents, including 18.33% neutrals, chose "strongly willing" and "willing"; as for the answers to the question: "To what extent are you willing to pay higher taxes for environmental protection," 52.78% of the respondents, including 18.85% neutrals, chose "strongly willing" and "willing".

Pecchenino [12] and Ono [121] to motivate the government to develop sustainable development policies and the residents to jointly participate in environmental governance, researches the "pollution co-governance" by the government, enterprises and residents, explores the institutional root of misallocation of environmental public goods and the theoretical mechanism thereof, and further uses the contract theory to put forward policy suggestions.

Chapter 3
Supply Status and Problems of Environmental Public Goods

Environmental issues are closely connected with our lives, and environmental quality has a direct impact on our physical health. Using the transnational panel database covering 89 countries of 1990–2010, Qi and Lu [125] found that environmental pollution became an important transmission channel by which the national health was affected in economic cycles; the health contribution from air pollution improvement accounted for 20% in the recession; the effects of economic cycles on health were significantly heterogeneous in terms of different ages, countries and genders.

3.1 Environmental Public Goods in Seriously Short Supply

The quality environment is an unswerving pursuit of mankind. Fresh air, clean water sources, superior living environment, and so on, can effectively improve our welfare level. With the rapid development of economy, roads, traffic, community-based public sports goods, and other infrastructure have been significantly improved, but the quality of the environment where humans live has been increasingly worse. Our living environment is being challenged by pollution problems.

3.1.1 Comparison of Air Pollution Between China and Overseas Countries

In 2016, among China's 338 cities at or above the prefecture level, only 84 cities met the ambient air quality standards, accounting for 24.9% of the total number of cities; 254 cities failed to meet the standards, accounting for 75.1%. In recent years, we have been concerned about air quality. One of the indexes for measuring air quality is air quality index (AQI). The index is the new standard for air quality

J. Yang, *Effective Supply Mechanism and Path of Environmental Public Goods in China*, Public Economy and Urban Governance in China,
https://doi.org/10.1007/978-981-99-2939-9_3

assessment released by China in March 2012, in which pollutant monitoring covers sulfur dioxide (SO_2), nitrogen dioxide (NO_2), PM_{10}, $PM_{2.5}$, carbon monoxide (CO), and ozone (O_3). Air pollution index (API) is divided into the following six categories: Level 1 represents excellent air, 0–50; Level 2 represents qualified air, 50–100; Level 3 represents mild pollution, 101–150; Level 4 represents moderate pollution, 151–200; Level 5 represents heavy pollution, 201–300; Level 6 represents severe pollution, above 300. When API is at Level 2, some pollutants may have a certain effect on the health of a tiny minority of unusually susceptible people, who should reduce outdoor activities. When Level 3 pollution occurs, susceptible people's symptoms will be aggravated mildly, and healthy people may have irritating symptoms. In this case, children, old people, as well as patients with heart disease or respiratory illnesses should reduce long-time and high-intensity exercises.

1. Physical situation of China's AQI

The average AQI values of 31 provinces, autonomous regions and municipalities in China from 2014 to 2017 are listed in Table 3.1. According to the data shown in the table, only the air quality of Hainan reached the excellent level in the four years; the air quality of Yunnan reached the excellent level in 2016 and 2017; other regions had no excellent air for four consecutive years. Moreover, Beijing, Hebei, Henan, Tianjin, Xinjiang, and so forth, had been in polluted air. All these showed that our environmental governance task was arduous.

2. Comparison of air pollutant $PM_{2.5}$ between China and overseas countries

According to environmental monitoring data: In 2005, 2010 and 2015, the average concentrations of particulate matter 2.5 ($PM_{2.5}$) were respectively 56.90 $\mu g/m^3$, 58.22 $\mu g/m^3$ and 58.38 $\mu g/m^3$ in China, far higher than the world's averages—41.78 $\mu g/m^3$, 41.84 $\mu g/m^3$ and 43.98 $\mu g/m^3$. However, the guideline values for PM_{10} and $PM_{2.5}$ set by the WHO were respectively 20 $\mu g/m^3$ and 10 $\mu g/m^3$ (WHO 2006), and the concentration of $PM_{2.5}$ was significantly higher in China, compared to either different regions or countries at different income levels. In Table 3.2, the concentrations of $PM_{2.5}$ in China were compared to those in other countries or regions from the perspectives of classification by income and region. It could be observed that China's air was seriously polluted. From the perspective of income, China's pollution was severer than the average level of low-income countries, and significantly severer than that of high-come countries. In 2005, the concentration of $PM_{2.5}$ was 15.65 $\mu g/m^3$ in high-income countries, and 56.90 $\mu g/m^3$ in China. After 10 years of pollution control, in 2015, the concentration of $PM_{2.5}$ was 58.38 $\mu g/m^3$ in China, and 16.65 $\mu g/m^3$ in high-income countries. From the perspective of region, China's pollution was severe. The data of 2015 suggested that China's air pollution was similar to that of the Middle East and North Africa (60.99 $\mu g/m^3$).

In Fig. 3.1, the $PM_{2.5}$ concentrations of 2005, 2010 and 2015 in different countries were compared according to income level. From 2005 to 2015, the average concentration of $PM_{2.5}$ in China not only was over five times that in high-income

Table 3.1 Average AQI values of 31 Provinces, autonomous regions and municipalities in China from 2014 to 2017

Region	2014	2015	2016	2017
Anhui	70.07	80.60	83.10	87.92
Beijing	127.20	123.43	113.86	107.84
Fujian	72.99	53.20	52.36	57.23
Gansu	132.50	83.51	80.58	81.01
Guangdong	68.57	59.67	58.40	61.01
Guangxi	98.58	63.80	59.18	58.43
Guizhou	112.05	56.40	56.00	50.71
Hainan	41.70	41.80	39.86	38.55
Hebei	135.06	115.64	110.52	113.55
Henan	93.49	115.25	113.05	109.10
Heilongjiang	82.49	68.03	58.67	61.21
Hubei	108.39	94.71	84.21	77.04
Hunan	105.28	88.05	77.02	72.28
Jilin	59.62	86.94	75.33	74.51
Jiangsu	106.05	92.21	84.94	84.88
Jiangxi	78.63	67.07	70.73	70.17
Liaoning	86.94	91.35	82.74	84.02
Inner Mongolia	64.49	77.64	72.43	74.98
Ningxia	111.77	89.74	89.59	90.07
Qinghai	83.47	78.80	73.42	69.87
Shandong	94.35	103.81	98.05	95.15
Shanxi	83.70	90.82	95.37	107.87
Shaanxi	98.73	89.28	97.25	92.78
Shanghai	83.15	89.68	80.89	82.61
Sichuan	113.44	77.16	77.08	72.39
Tianjin	120.58	103.69	104.08	110.37
Tibet	73.30	60.79	62.91	61.21
Xinjiang	135.63	104.78	110.25	106.01
Yunnan	78.20	51.21	48.71	49.46
Zhejiang	90.66	83.22	70.45	71.53
Chongqing	95.86	82.20	81.03	77.12
National Average	94.53	83.22	79.79	79.31

Data source Calculated according to the data from the Ministry of Ecology and Environment (formerly Ministry of Environmental Protection)
Note As the information in the database is AQI value captured every day, the data of each province in the table is obtained by averaging the sum of data over 365 days

Table 3.2 Comparison of $PM_{2.5}$ concentrations between China and other Countries or Regions

Country or region		2005	2010	2015
By income	China	56.90	58.22	58.38
	High-income countries	15.65	16.56	16.65
	Upper-middle income countries	41.79	42.35	42.16
	Middle-income countries	47.84	47.63	50.28
	Lower-middle income countries	47.09	46.83	49.08
	Low-income countries	39.45	39.36	38.72
By region	East Asia and Pacific	43.59	43.30	44.16
	Europe and Central Asia	16.75	18.26	19.38
	EU	14.32	14.72	15.34
	Latin America and the Caribbean	21.82	17.21	17.93
	Middle East and North Africa	49.60	61.88	60.99
	North America	10.19	8.48	8.31
	Sub-Saharan Africa	42.04	39.69	36.06
	World average	41.78	41.84	43.98

Data source World Bank Database
Unit $\mu g/m^3$

countries, but also was 50% higher than that in low-income countries. Certainly, the figure showed that the average concentrations of $PM_{2.5}$ in middle-income countries and lower-middle income countries were on the rise in 2015, but the average concentration of $PM_{2.5}$ in China remained far higher than that in other income countries in the same period.

In Fig. 3.2, the comparison was conducted between regions. North America and the EU had the best air quality, but China was still one of the regions poor in air quality (measured according to the average concentration of $PM_{2.5}$); in 2010, the concentration was merely 3.66% (2.61% lower than that in 2015) lower than that in the Middle East and North Africa. Unexpectedly, China's air pollution was similar to that of the Middle East and North Africa.

Seen from both Table 3.3 and Fig. 3.3, the monthly average concentration of $PM_{2.5}$ in China showed a U-shaped trend in accordance with either of the statistical standards; the pollution of the Beijing-Tianjin-Hebei region had been severer than that of the Yangtze River Delta and Pearl River Delta regions, and the monthly average concentration of the former was higher than that of 74 cities. In 2014, among China's 161 cities monitored pursuant to the *Ambient Air Quality Standards* (GB 3095–2012), only 9.9% of them met the air quality standards, and 90.1% failed to meet the standards. In 2016, among China's 338 cities monitored by the Ministry of Environmental Protection (present-day Ministry of Ecology and Environment), only 24.9% of them met the air quality standards, and 75.1% failed to meet the standards.

In 2017, among the 338 cities at or above the prefecture level in China, 239 cities failed to meet the ambient air quality standards, accounting for 70.7%, with

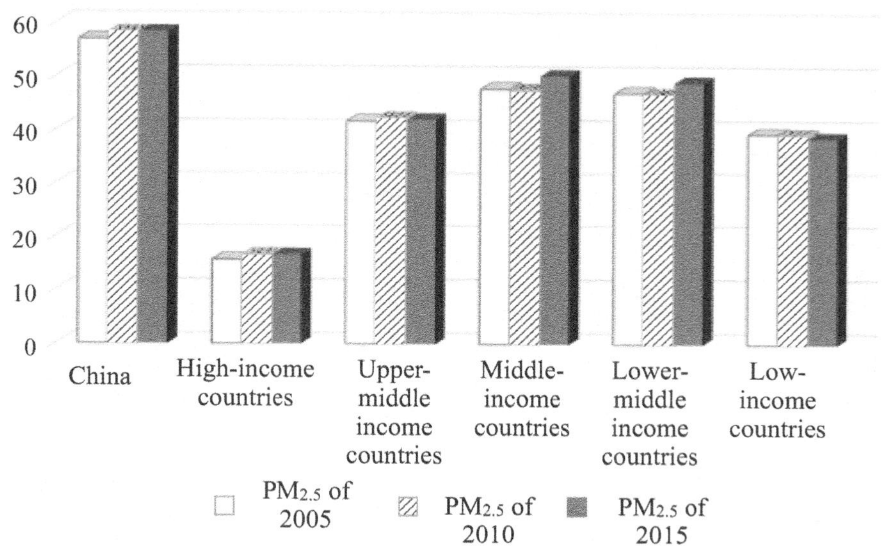

Fig. 3.1 International comparison of $PM_{2.5}$ concentrations: countries at different income levels. *Data source* World Bank Database

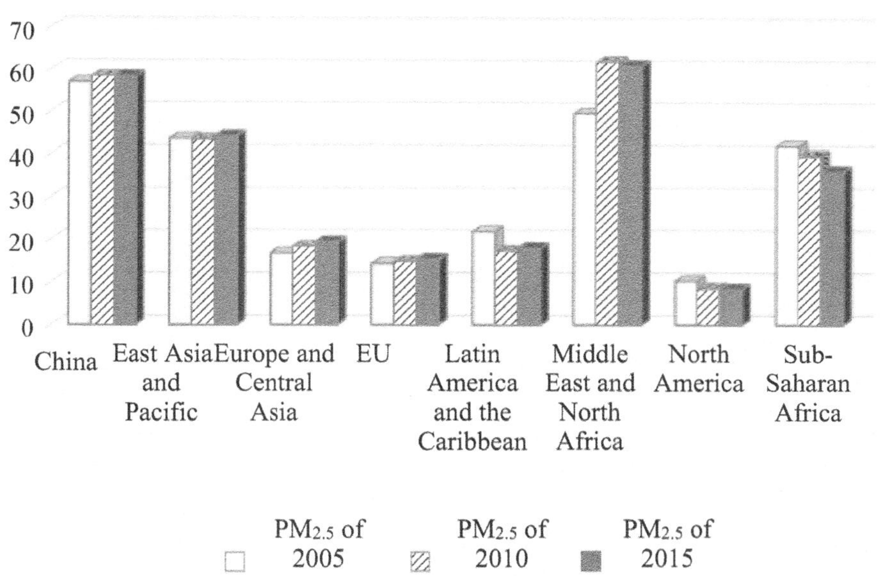

Fig. 3.2 International comparison of $PM_{2.5}$ concentrations: different regions. *Data source* World Bank Database

Table 3.3 Comparison of monthly average concentrations of $PM_{2.5}$ among 74 major cities and three regions in China

Time	74 cities	Beijing-Tianjin-Hebei region	Yangtze River Delta	Pearl River Delta
January 2017	81	128	60	51
February 2017	67	92	63	41
March 2017	51	63	50	39
April 2017	44	55	46	33
May 2017	40	54	38	33
June 2017	32	47	34	15
July 2017	30	50	28	19
August 2017	27	39	26	20
September 2017	34	52	30	30
October 2017	41	61	33	34
November 2017	53	60	50	42
December 2017	68	73	70	55
January 2018	67	70	72	53
February 2018	57	73	55	43
March 2018	53	80	48	35
April 2018	44	52	47	34
May 2018	35	44	38	21

Data source China National Environmental Monitoring Centre
Unit μg/m^3

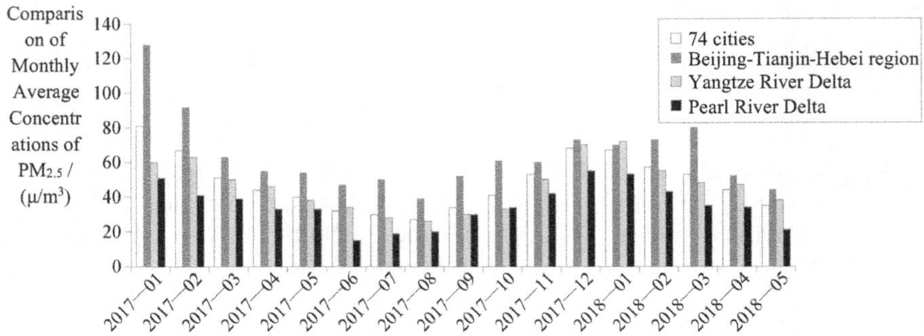

Fig. 3.3 Comparison of monthly average concentrations of $PM_{2.5}$ among 74 major cities and three regions in China. *Data source* China National Environmental Monitoring Centre

the average proportion of days below the standards reaching 22.0%, and the average proportion of days with good air quality reaching 78.0%. The 338 cities experienced 2311 days of heavy pollution, and 802 days of severe pollution, and the number of days with $PM_{2.5}$ as the major pollutant accounted for 74.2% of the days with heavy pollution and above. The average proportion of days with good air quality in 74 cities was 72.7%, declining by 1.5% over 2016; the average proportion of days below the standards was 27.3%. In 2018, the number of cities failing to meet air quality standards declined to 217, with the average proportion of days below the standards reaching 20.7%; the 338 cities experienced 1899 days of heavy pollution, decreasing by 412 days over 2017, but underwent 822 days of severe pollution.

3.1.2 Water Pollution Status of China

In 2013, the results of the first nationwide large-scale study published by the Ministry of Environmental Protection suggested that China's surface water was mildly polluted on the whole, and some city river sections were seriously polluted. Among China's 4778 groundwater monitoring sites, the proportion of sites with poor and very poor water quality was nearly 60%. 17 out of 31 large freshwater lakes were moderately or mildly polluted. Among 4896 groundwater monitoring sites, 10.8% had excellent water quality, 25.9% had good water quality, 1.8% had fair water quality, 45.4% had poor water quality, and 16.1% had very poor water quality.

1. Total wastewater discharge rising year by year

According to Fig. 3.4, China's total wastewater discharge was approximately 55.685 billion tons in 2007, and progressively increased at a rate of around 3% every year. From 2010 to 2011, the total wastewater discharge had a sharp increase, rising by 6.8% over the previous year. According to the annual report published by the Ministry of Environmental Protection, in 2015, the total national wastewater discharge was 73.53 billion tons, an increase of 2.7% over 2014. The total industrial wastewater discharge was 19.95 billion tons, a decrease of 2.8% over 2014; accounted for 27.1% of the total wastewater discharge, decreasing by 1.6% over 2014. The urban domestic sewage discharge was 53.52 billion tons, an increase of 4.9% over 2014; accounted for 72.8% of the total wastewater discharge, increasing by 1.5% over 2014.[1]

It is usually accepted that with the acceleration of economic growth, the wastewater discharge will increase accordingly. Thus, this book associates the growth rate of wastewater discharge with the economic growth rate for comparison. Seen from Table 3.4, in 2011 when the wastewater discharge grew relatively fast, the economic growth rate was 9.2%, with no significantly sharp rise.

Additionally, seen from Fig. 3.5, in 2010 when the economic growth rate was the highest, wastewater discharge growth was not the highest. It followed that China's

[1] The data comes from the *2016 Report on the State of the Environment in China.*

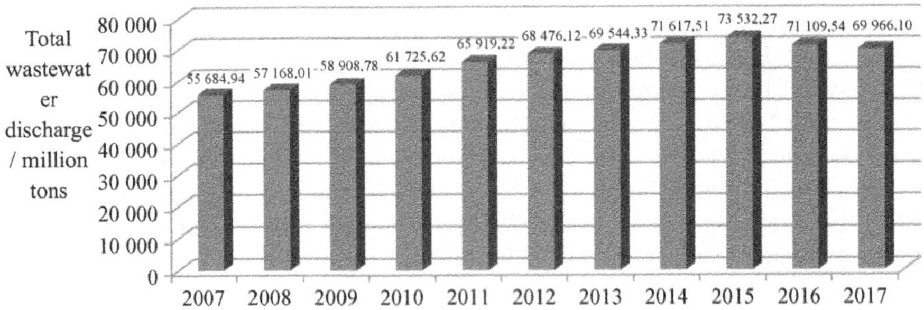

Fig. 3.4 Total wastewater discharge in China from 2007 to 2017. *Data source* China Statistical Yearbooks 2007–2017

Table 3.4 Wastewater discharge in China from 2007 to 2017

Year	Total wastewater discharge (million tons)	Growth rate of wastewater discharge (%)	GDP growth rate (%)
2007	55,684.94	–	11.9
2008	57,168.01	2.66	9.0
2009	58,908.78	3.05	9.1
2010	61,725.62	4.78	10.3
2011	65,919.22	6.79	9.2
2012	68,476.12	3.88	7.8
2013	69,544.33	1.56	7.7
2014	71,617.51	2.98	7.4
2015	73,532.27	2.67	6.9
2016	71,109.54	− 3.29	6.7
2017	69,966.10	− 1.61	6.9

Data source Calculated according to the data of China Statistical Yearbooks 2007–2017

wastewater discharge did not match the variation trend of GDP. In consideration of the total wastewater discharge in Fig. 3.4, it was observed that China's total wastewater discharge showed a rising trend from 2007 to 2015. It was gratifying that the total wastewater discharge showed a decline trend from 2016.

2. Aggravated water pollution

The China Statistical Yearbook makes statistics on the number of occurrences for various environmental emergencies in China every year.[2] In Table 3.5, we review

[2] In 2010 and previous years, the total number of occurrences of various environmental emergencies, including water, air, solid waste, noise and vibration, as well as other pollution, was recorded in China Statistical Yearbooks; after 2010, the total number of occurrences thereof was still recorded, but the subcategories were changed to extraordinarily severe, severe, comparatively severe, ordinary and unrated environmental emergencies.

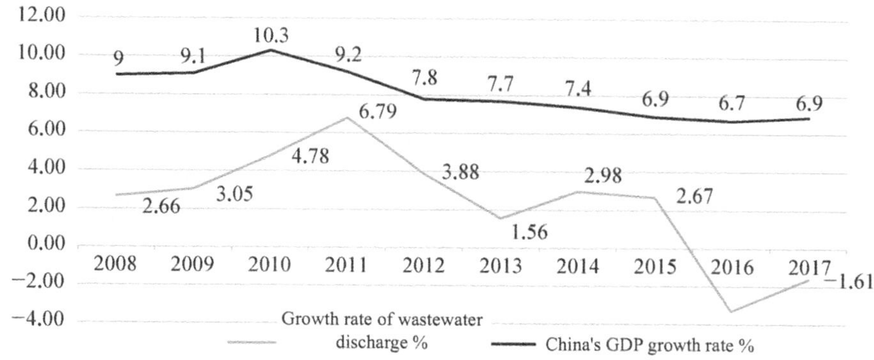

Fig. 3.5 Comparison between growth rate of wastewater discharge and economic growth rate. *Data source* Calculated according to the data of China Statistical Yearbooks 2007–2017

the number of occurrences of various environmental emergencies in China from 2000 to 2010. Water pollution is particularly noteworthy. For four consecutive years from 2001 to 2004, and in 2006, water pollution emergencies accounted for over 50% of various environmental emergencies in China, and even 59.5% in 2001. In 2016, with the groundwater system as the unit, among 6124 groundwater quality monitoring sites focusing on shallow groundwater dominated by phreatic water, and medium and deep groundwater dominated by confined water, the monitoring sites with excellent, good, fair, poor and very poor water quality respectively accounted for 10.1%, 25.4%, 4.4%, 45.4% and 14.7%, which meant that 60.1% of monitoring sites were "poor" in water quality.

Upon research, Zhang [126] pointed out that China's river, lake, and offshore pollution showed a rising trend on the whole; reservoir and lake pollution grew faster than or at the same rate with the economic aggregate in the same period. In his view, China paid a huge price in water pollution for economic growth.

See Fig. 3.6 for the data of major pollutant (chemical oxygen demand (COD) and ammonia nitrogen) emissions in discharged wastewater.

From 2001 to 2010, both COD and ammonia nitrogen emissions showed a slightly downward trend, but after 2011, they both rose sharply, and then showed a downward trend. To compare the variation trends of pollutants in wastewater during economic growth in a better fashion, we have searched the ammonia nitrogen emission, COD emission and GDP of 31 provinces, autonomous regions and municipalities. Besides, according to the typical classification of East, Central and West China: East China includes Beijing, Tianjin, Hebei, Liaoning, Shanghai, Jiangsu, Zhejiang, Fujian, Shandong, Guangdong and Hainan; Central China includes Shanxi, Jilin, Heilongjiang, Anhui, Jiangxi, Henan, Hubei and Hunan; West China includes Inner Mongolia, Guangxi, Chongqing, Sichuan, Guizhou, Yunnan, Tibet, Shaanxi, Gansu, Qinghai, Ningxia and Xinjiang; we have calculated the wastewater pollutant emissions per unit of output in all regions, i.e. the amount of pollutants (ton) discharged per 100 million Yuan output. See Table 3.6 for the specific correlation data from 2010 to 2015.

Table 3.5 Various environmental emergencies in China (2000–2001)

Year	Total number of occurrences	Water pollution/ number of occurrences	Air pollution/ number of occurrences	Solid waste pollution/ number of occurrences	Noise and vibration/ number of occurrences	Other pollution/ number of occurrences	Proportion of water pollution (%)
2000	2411	1138	864	103	266	40	47.20
2001	1842	1096	576	39	80	51	59.50
2002	1921	1097	597	109	97	21	57.11
2003	1843	1042	654	56	50	41	56.54
2004	1441	753	569	47	36	36	52.26
2005	1406	693	538	48	63	64	49.29
2006	842	482	232	45	6	77	57.24
2007	462	178	134	58	7	85	38.53
2008	474	198	141	45	–	90	41.77
2009	418	116	130	55	–	117	27.75
2010	420	135	157	35	1	92	32.14

Data source China Statistical Yearbooks 2000–2010
Note The proportion of water pollution was calculated based on original data

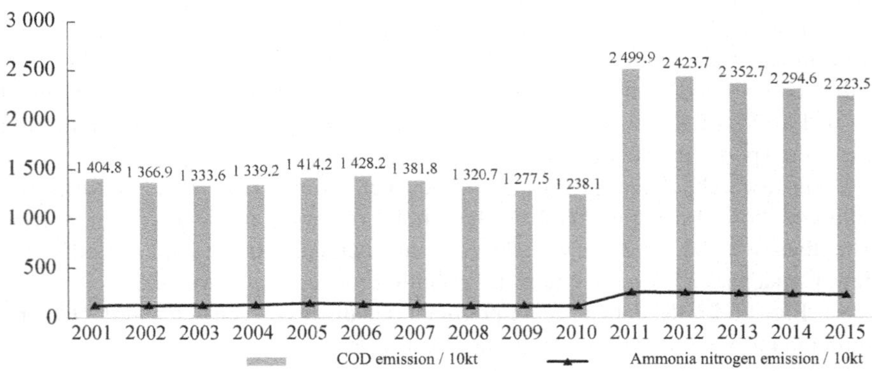

Fig. 3.6 COD and ammonia nitrogen emissions in wastewater discharged in China[3]

In Fig. 3.7, the pollutant emissions in wastewater (per unit output) of 2010 and 2015 in East, Central and West China were compared. According to the data, only the ammonia nitrogen and COD emissions of West China both declined, and the COD emission of 2015 declined by 12.95% compared to that of 2010; the two indicators of East and Central China both rose, the ammonia nitrogen emission (per unit of

[3] The data comes from *2001 to 2015 China Statistical Yearbooks on Environment*. Since 2011, agricultural pollutant emissions have been added to environmental statistics, where agricultural sources include pollutants from planting, aquaculture, as well as livestock and poultry.

Table 3.6 Wastewater pollution indicators in East, Central and Western China

Region	Ammonia nitrogen emission/(ton/100 million Yuan)			COD emission/(ton/100 million Yuan)		
	2010	2015	Change proportion (%)	2010	2015	Change proportion (%)
East China	2.10	2.62	24.76	20.98	21.01	0.14
Central China	3.88	4.39	13.14	38.86	45.98	18.32
West China	4.39	4.06	− 7.52	49.41	43.01	− 12.95

Data source The pollutant emissions in wastewater were obtained from the data of China Statistical Yearbooks (2010, 2015); GDP was obtained from the website of the National Bureau of Statistics; ammonia nitrogen and COD emissions were obtained by the division of regional total pollutant emissions by GDP of the year

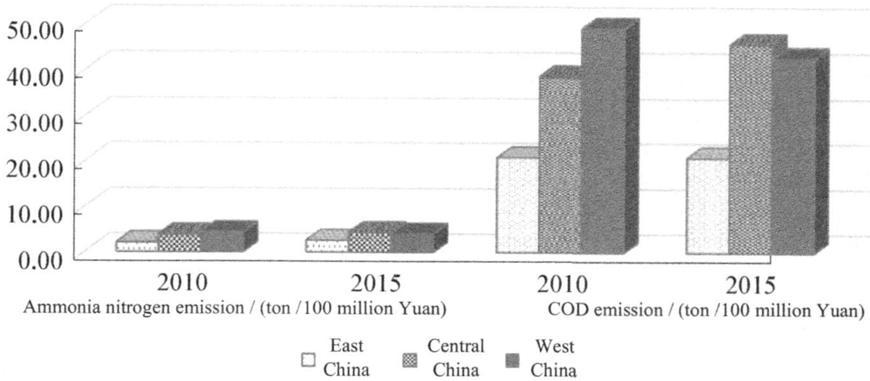

Fig. 3.7 Comparative analysis on major pollutants in wastewater of China. *Data source* The pollutant emissions in wastewater were obtained from the data of China Statistical Yearbook; GDP was obtained from the website of the National Bureau of Statistics; ammonia nitrogen and COD emissions were obtained by the division of regional total pollutant emissions by GDP of the year

output) of East China rose by 24.76%, and the COD emission (per unit of output) of Central China rose by 32% compared to that five years ago.

3.1.3 Waste Pollution Status of China

1. Industrial solid waste pollution

In 2015, China produced 3.27 billion tons of general industrial solid waste, increasing by 0.4% over 2014. The industries producing the largest amount of solid waste were respectively ferrous metal mining and dressing, electricity and heat production and supply, ferrous metal smelting and pressing, coal mining and washing, nonferrous metal mining and dressing, chemical raw material and product manufacturing, as

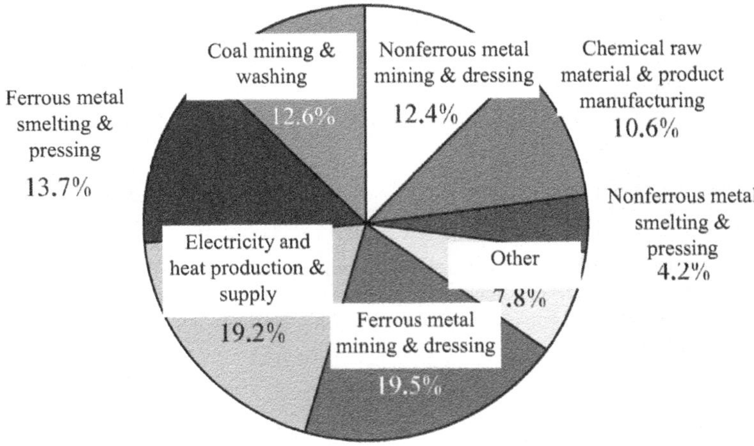

Fig. 3.8 China's key industries for industrial solid waste pollution in 2015. *Data source* 2015 China Statistical Yearbook on Environment

well as nonferrous metal smelting and pressing. The above industries respectively accounted for 19.5%, 19.2%, 13.7%, 12.6%, 12.4%, 10.6% and 4.2% of the solid waste produced by industrial enterprises under intensive investigation, and the sum of solid waste produced by the seven industries accounted for 92.2% of the total solid waste produced (as shown in Fig. 3.8).

Another industrial solid pollutant is the amount of industrial solid hazardous waste produced, mainly including waste alkali, waste acid, asbestos waste, nonferrous metal smelting waste, inorganic cyanide, and waste mineral oil. In 2015, China produced 397,600 tons of industrial solid hazardous waste, increasing by 9.4% over 2014 (as shown in Fig. 3.9).

2. Residents' household waste pollution

With the development of economy and the improvement of people's living standards, the waste problem in China has become increasingly prominent (Fig. 3.10).

It is reported that two thirds of the 668 cities in China are surrounded by waste. Waste will bring a series of pollution problems: the stacking or simple landfill of household waste may easily cause air pollution; the leachate from rainwater in waste and water contained therein may lead to water pollution after flowing into surrounding surface water; plastic bags, scrap metal, waste glass and other bio-refractory toxic substances among waste may be left in soil, resulting in soil pollution; lots of pathogenic microorganisms exist in waste, severely affecting our health. The household waste produced by residents is too much to be fully buried or incinerated, and is a severe hazard to society.

In addition to air, water and waste pollution, there is soil, noise and other pollution. Besides, importance must be attached to such environmental problems as insufficient

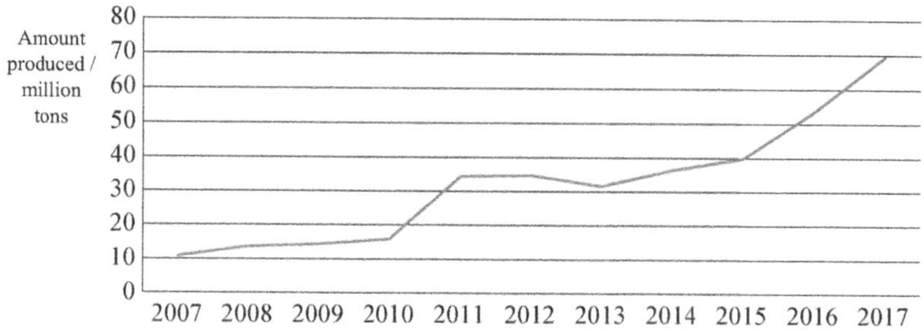

Year	2007	2008	2009	2010	2011	2012	2013	2014	2015	2016	2017
Amount of industrial solid hazardous waste produced / million tons	10.79	13.57	14.30	15.87	34.31	34.65	31.57	36.34	39.76	53.47	69.37

Fig. 3.9 Amount of industrial solid hazardous waste produced in China from 2007 to 2017. *Data source* Data from the website of the National Bureau of Statistics from 2007 to 2017

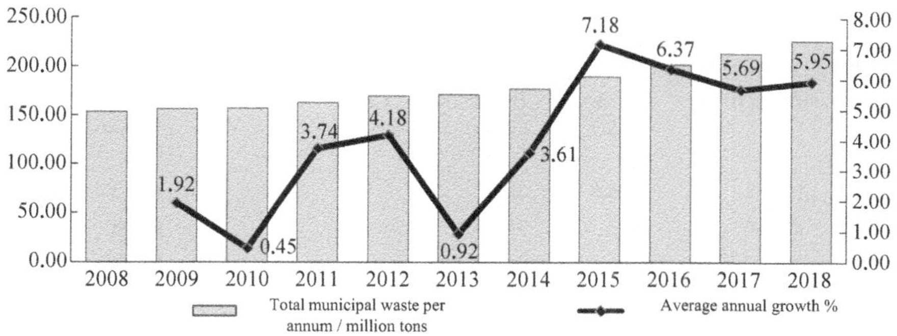

Fig. 3.10 Amount of municipal solid waste collected in China every year. *Data source* Calculated based on related data from the website of National Bureau of Statistics

green space, destruction of forest vegetation, shortage of fresh water, desertification and reduction of animals and plants.

Empirical researches suggest that environmental pollution has significantly damaged social well-being, not only lowered our work efficiency, human capital and residents' happiness, but also severely affected our health, led to the rise of the death rate of lung cancer and the incidence of cardiovascular diseases, and shortened residents' life expectancy [2].

3.2 Analysis on Sources of Environmental Pollution

Faced with such harsh environmental pollution, we have to wonder: What are the main sources of pollution? Which individuals or enterprises bring such pollution? Which industries are the focuses of pollution control? This section mainly analyzes the structure of each pollution source.

3.2.1 *Severe Industrial Pollution*

The vital indicators of air pollution include SO_2 and nitrogen oxide (NO_x) emission, as well as soot (dust) emission. The data in Tables 3.7, 3.8 and 3.9 clearly shows that no matter which indicator is used for comparison, the pollution from industrial production is severer than household waste pollution caused by residents. In the total emissions of SO_2, the proportion of industrial emission had been over 80% over the years, and even reached 90.95% and 90.27% in 2011 and 2012.

Table 3.7 Total Emissions of SO_2 in China from 2001 to 2015

Year	Total (10 kt)	Industrial (10 kt)	Household (10 kt)	Centralized① (10 kt)	Proportion of industrial SO_2 (%)
2001	1947.8	1566.6	381.2	–	80.43
2002	1926.6	1562	364.6	–	81.08
2003	2158.7	1791.4	367.3	–	82.99
2004	2254.9	1891.4	363.5	–	83.88
2005	2549.3	2168.4	380.9	–	85.06
2006	2588.8	2237.6	351.2	–	86.43
2007	2468.1	2140	328.1	–	86.71
2008	2321.2	1991.3	329.9	–	85.79
2009	2214.4	1865.9	348.5	–	84.26
2010	2185.1	1864.4	320.7	–	85.32
2011	2217.9	2017.2	200.4	0.3	90.95
2012	2117.7	1911.7	205.7	0.3	90.27
2013	2043.9	1835.2	208.5	0.2	89.79
2014	1974.5	1740.4	233.9	0.2	88.14
2015	1859.12	1556.7	296.9	–	83.98

Data source Data was obtained from 2001 to 2015 China Statistical Yearbooks on Environment, and the data of the last column was obtained by calculation

① Emission from centralized pollution control facilities refers to the emission of waste leachate/wastewater and pollutants from household waste treatment plants and centralized hazardous waste (medical waste) treatment plants (signifying the same in Tables 3.8 and 3.9)

Table 3.8 Total emissions of NO_x in China from 2006 to 2015[4]

Year	Total/ 10 kt	Industrial/ 10 kt	Household/ 10 kt	Motor vehicles/10 kt	Centralized/ 10 kt	Proportion of industrial NO_x (%)
2006	1523.8	1136	387.8	–	–	74.55
2007	1643.4	1261.3	382.1	–	–	76.75
2008	1624.5	1250.5	374	–	–	76.98
2009	1692.7	1284.8	407.9	–	–	75.90
2010	1852.4	1465.6	386.8	–	–	79.12
2011	2404.3	1729.7	36.6	637.6	0.3	71.94
2012	2337.8	1658.1	39.3	640	0.4	70.93
2013	2227.4	1545.6	40.8	640.6	0.4	69.39
2014	2078.0	1404.8	45.1	627.8	0.3	67.60
2015	1851.02	1180.9	65.1	585.9	–	64.46

Data source Data was obtained from 2006 to 2015 China Statistical Yearbooks on Environment, and the data of the last column was obtained by calculation

Table 3.9 Total emissions of soot (dust) in China from 2006 to 2015

Year	Total/ 10 kt	Industrial/ 10 kt	Household/ 10 kt	Motor vehicles/ 10 kt	Centralized/ 10 kt	Proportion of industrial soot (dust) (%)
2006	1088.5	864.5	224	–	–	79.40
2007	987.1	771.1	216	–	–	78.16
2008	901.7	670.7	231	–	–	74.39
2009	847.4	604.4	243	–	–	71.30
2010	829.2	603.2	226	–	–	72.75
2011	1278.8	1100.9	115	62.9	0.2	86.09
2012	1234.3	1029.3	143	62.1	0.2	83.39
2013	1278.14	1094.6	124	59.4	0.2	85.64
2014	1740.8	1456.1	227	57.4	0.2	83.65
2015	1538.0	1232.6	250	55.5	–	80.14

Data source Data was obtained from 2006 to 2015 China Statistical Yearbooks on Environment, and the data of the last column was obtained by calculation

The industrial proportion of NO_x emission was lower than that of SO_2 emission. Seen from the data in Table 3.8, the industrial proportion of NO_x emission was above 60% every year, and maintained above 70% before 2012. In 2010, NO_x emission reached its highest industrial proportion, 79.12%. There was a gratifying trend that

[4] China began to make statistics on the emission of NO_x in 2006, wherein the household emission included the NO_x emitted by the traffic source. Since 2011, the emission of motor vehicle exhaust pollutant had been separated from that of the household source for separate statistics.

it began to decline in 2011, and the industrial proportion of this indicator declined to 64.46% in 2015.

Table 3.9 lists the total emissions of soot (dust) in China in recent years. Since 2006, the industrial emission of soot (dust) had fluctuated significantly, and showed a sharply upward trend after 2011. Despite the variations, according to statistical data, the industrial emission of soot (dust) accounted for more than 70% of the total emissions every year, and even reached 85.64% in 2013.

3.2.2 Progressive Increase of Domestic Pollution Caused by Residents Year by Year

In light of the data regarding the industrial emission of various waste gases, soot, and so on, we may consider that enterprises are definitely the major source of environmental pollution. Some even say that all pollution sources come from enterprises. Is this view objective? Let's turn to residents' domestic pollutant discharge.

1. Progressive rise of residents' domestic sewage discharge

Seen from the statistical data, the discharge of wastewater in China had shown an upward tendency. The total discharge of wastewater was 55.685 billion tons in 2007, and increased to 68.476 billion tons in 2012, an increase of 22.97%; even reached 73.532 billion tons in 2015, increasing by 32.05% compared to eight years ago.

Wastewater discharge comprises two parts: domestic and industrial sewage. It was gratifying that the discharge of industrial wastewater in China showed a significantly downward trend (as shown in Fig. 3.11), declining by 19.12% in 2015 compared to 2007. Domestic sewage caused the continuous rise of China's wastewater discharge in recent years. From 2007 to 2015, residents' sewage discharge remained stubbornly high, and showed a rising trend year by year. During the eight years, the discharge of domestic sewage rose by 72.54%, from 31.020 billion tons to 53.520 billion tons.

As regards the industrial and domestic sources of wastewater discharge in 31 provinces/municipalities/autonomous regions of China, is the discharge of industrial wastewater lower than that of domestic wastewater, similar to the overall trend? Or is the opposite the case? Or does the case vary with regions? Fig. 3.12 lists the wastewater discharge conditions of different parts of China in 2015.

To our amazement, with regard to the proportion of sewage discharge in these provinces, autonomous regions and municipalities, the discharge of domestic sewage was far higher than that of industrial sewage. The top three regions in the discharge of urban domestic sewage were Guangdong, Jiangsu and Shandong, which respectively accounted for 14.0%, 7.7% and 7.0% of the total discharge of urban domestic sewage nationwide. In Guangdong, the domestic wastewater discharge was nearly six times that of industrial wastewater; in Jiangsu and Shandong, the domestic wastewater discharge was three times that of industrial wastewater.

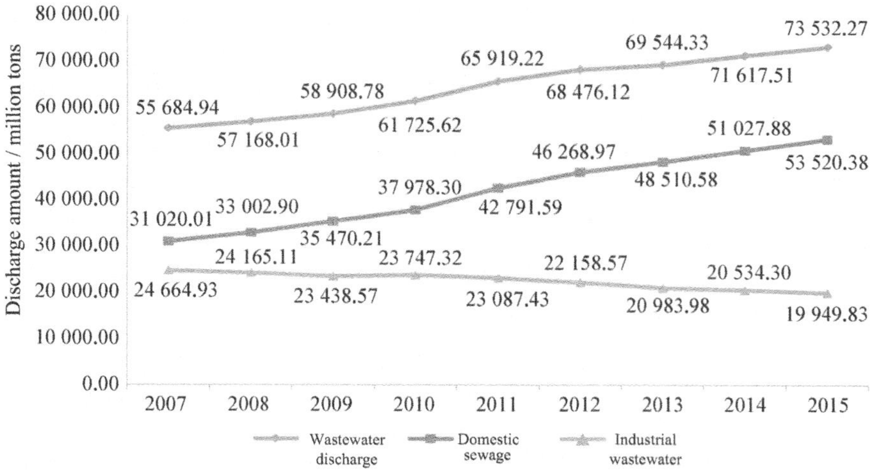

Fig. 3.11 Wastewater discharge of China from 2007 to 2015. *Data source* Obtained from the data of the Ministry of Environmental Protection

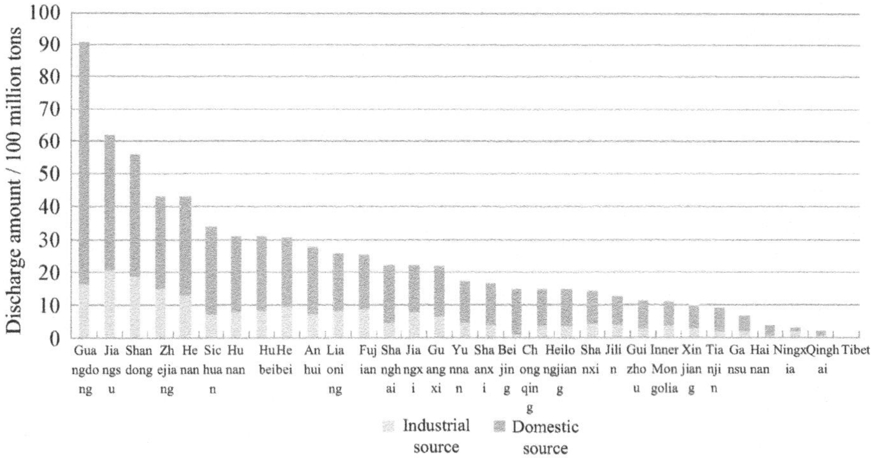

Fig. 3.12 Wastewater discharge in different parts of China in 2015. *Data source* Obtained from 2015 China Statistical Yearbook on Environment

2. Severe household waste pollution caused by residents

Residents may subjectively consider that the pollution in life mainly results from the discharge of three wastes by enterprises, but in fact, every living being on the earth is one of the pollution sources.

In Table 3.10 and Fig. 3.13, we could observe that China's industrial solid waste grew fast from 2009 to 2011, which increased by 18.14% in 2010 over the previous year, and rose by 35.39% in 2011 over 2010. The growth rate of household solid fluctuated up and down before 2013, but continuously rose thereafter, which rose by 3.60% in 2014 over the previous year while the growth rate of industrial solid waste declined by 1.58% in the same period. In 2015, the growth rate of household waste rose by 7.18% while that of industrial waste only rose by 0.45%. The development trend suggested that industrial solid waste pollution slowed down, but the household waste pollution caused by residents continuously rose together with the growth rate of domestic water pollution. Therefore, China should attach greater importance to the treatment of domestic pollution sources from residents.

3. Severe air pollution arising from residents' travel

It is universally known that traffic congestion is one of the travel difficulties in most cities of China. During rush hours, urban roads are usually swarming with private vehicles. Seemingly, individual residents or families obtain benefits. However, after making statistics on the air pollution from car exhaust emission, we could easily find that residents' travel has brought severe air pollution indeed.

According to Fig. 3.14, apart from Beijing, the number of vehicles privately owned per 100 persons in other provinces, autonomous regions and municipalities showed a sharp rise trend in 2015 compared to that in 2008. In Jiangxi and Gansu, it respectively rose to 4.5 and 4.2 times of the previous number; in Chongqing, Guangxi, Qinghai and Jilin, it was over 3 times of the previous number; in other

Table 3.10 Growth status of industrial solid waste and residents' household waste in China

Year	2009	2010	2011	2012	2013	2014	2015
Growth rate of household waste (%)	1.92	0.45	3.73	4.18	0.93	3.60	7.18
Growth rate of industrial solid waste (%)	7.27	18.14	35.39	1.93	− 0.50	− 1.58	0.45

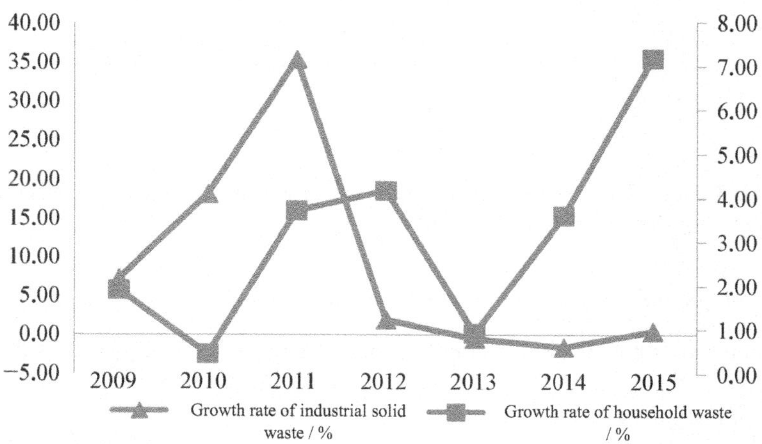

Fig. 3.13 Comparison between industrial solid waste and residents' household waste in China. *Data source* Obtained from the data of the former Ministry of Environmental Protection

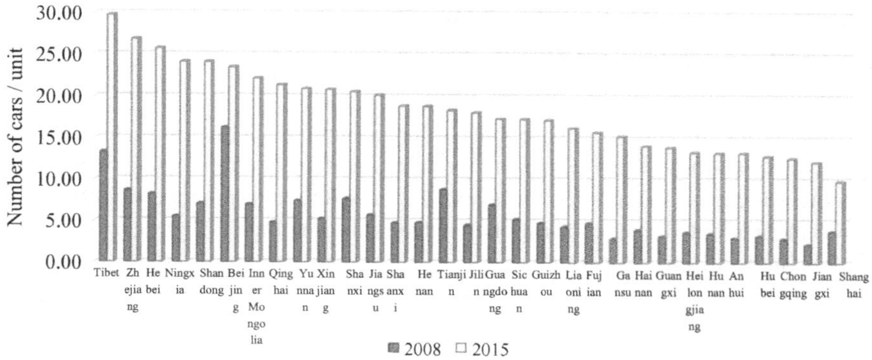

Fig. 3.14 Number of cars privately owned per 100 persons among Urban Residents in China. *Data source* Calculated according to the data from China Statistical Yearbooks and the Ministry of Environmental Protection

provinces, autonomous regions and municipalities, it was approximately 2 times of the previous number. The possession of private cars in Beijing was high in 2008, with 16.24 cars per 100 persons, so the indicator therein only rose by 44% in 2015 compared to that in 2008. In 2015, Tibet had the highest number of vehicles privately owned per 100 persons – 29.71, followed successively by Zhejiang, with 26.8, and Hebei, with 25.68. The continuous rise in the number of cars privately owned by residents brought huge environmental pressure.

Moreover, residents' food, clothing, accommodation and transportation have produced a great deal of pollution intentionally and unintentionally. In addition to the aforesaid air pollution from vehicle exhaust emission, water pollution caused by domestic water discharge, and pollution resulting from household waste, there are a lot of other pollution sources inconvenient for detailed description, e.g. moped, taxi exhaust emission, scooter exhaust emission, pollution from range hoods, and noise pollution. In North China, there is air pollution from coal combustion, as well as a series of pollution sources like straw burning and insecticide spraying in rural areas. Some pollution sources are too fragmented to be counted, thus the aforesaid proportion of pollution resulting from residents is far less than that of pollution actually produced.

It follows that the demand side (residents) usually with a strong demand for clean environment is the source of all sorts of pollution as well. If we can change the notion of the demand side (residents) and turn them into the supplier of quality environment, environmental governance will certainly develop as we expect. Do residents have such environmental awareness? Are they willing to bear part of pollution control fees? How should we motivate residents to participate? These are significant issues to be discussed subsequently.

3.3 Reason Analysis on Problems in Supply of Environmental Public Goods

In recent years, the undersupply problem of environmental public goods has become increasingly serious. The government's role orientation in public goods supply has become a focus of social attention again. Faced with environmental pollution, the government adopted a series of rationalized measures including closing, suspension, merging and shifting, as well as economic punishment measures, but the effect was insignificant. According to our analysis, it has the following reasons.

3.3.1 Administrative Intervention Overvalued and Market Regulation Undervalued in Environmental Public Goods Supply Policies

Since the promulgation and implementation of the first *Environmental Protection Law of the People's Republic of China* in 1989, to solve the increasingly intensified environmental problems in rapid socioeconomic development, the National People's Congress and its standing committee, the local people's congresses and their standing committees, the State Council and its subordinate ministries have constantly developed and promulgated new laws, regulations, policies and rules on environmental protection, with the chief aim to protect the environment. There are too many policies and regulations, which cannot be fully described herein. Table 3.11 only lists China's environmental protection policies, laws, regulations, etc. adopted in the past decade.

These policies promulgated by China are connected with not only administrative orders and financial laws and regulations, but also regulations implemented by financial means, and cover a wide range of areas, including water, electricity, coal, noise, air, and disposal of waste electronic products.

Additionally, at the eighth session of the Standing Committee of the 12th National People's Congress on April 24, 2014, the *Environmental Protection Law of the People's Republic of China* was revised, and a decision was made to implement the law as from January 1, 2015. At the 28th session of the Standing Committee of the 12th National People's Congress on June 27, 2017, the *Water Pollution Prevention and Control Law of the People's Republic of China* was revised, and a decision was made to implement the law as from January 1, 2018.

The above policies, laws and regulations were independently or jointly released by the former Ministry of Environmental Protection, the NDRC, the National Energy Administration, the Ministry of Finance, the former GAQSIQ, the former CBRC, and so on. Most of them were jointly issued by ministries and commissions, which fully demonstrated that environmental protection was a big issue influencing all walks of life.

The *Opinions on Comprehensively Strengthening Ecological and Environmental Protection and Resolutely Fighting the Uphill Battle for the Prevention and Control of*

Table 3.11 China's environmental protection policies, laws, regulations, etc. adopted in the past decade

Date	Environmental protection policies, laws, regulations, etc.	Issuing authority
2007.1.20	*Several Opinions on Accelerating the Closedown of Small Firepower Generators*	The National Development and Reform Commission of the People's Republic of China (NDRC), and the Office of the National Energy Leading Group (Energy Office)
2007.5.11	*Interim Measures for the Administration of the Special Fund Project of the Central Treasury for Reduction of Major Pollutant Discharge*	The State Environmental Protection Administration (SEPA), and the Ministry of Finance of the People's Republic of China (Ministry of Finance)
2007.7.12	*Opinions on Implementing Environmental Protection Policies and Rules and Preventing Credit Risks*	The SEPA, the People's Bank of China, and the China Banking Regulatory Commission (CBRC)
2007.7.30	*Notice of the General Office of the State Council on the Establishment of the System of Compulsory Government Procurement of Energy-Saving Products*	The General Office of the State Council
2007.12.4	*Guiding Opinions on Environmental Pollution Liability Insurance*	The SEPA, and the China Insurance Regulatory Commission (CIRC)
2008.2.28	*Water Pollution Prevention and Control Law of the People's Republic of China*	The Standing Committee of the National People's Congress of the People's Republic of China (NPC Standing Committee), and the Order of the President of the People's Republic of China
2008.8.19	*Emission Standard for Industrial Enterprises Noise at Boundary*	The Ministry of Environmental Protection of the People's Republic of China (Ministry of Environmental Protection), and the General Administration of Quality Supervision, Inspection and Quarantine (GAQSIQ)
2007.10.28	*Energy Conservation Law of the People's Republic of China*	The NPC Standing Committee
2008.8.29	*Circular Economy Promotion Law of the People's Republic of China*	The NPC Standing Committee, and the Order of the President of the People's Republic of China
2009.6.28	*Measures for the Implementation of Home Appliance Trade-In*	Seven ministries and commissions, including the Ministry of Commerce of the People's Republic of China (Ministry of Commerce), and the Ministry of Finance

(continued)

Table 3.11 (continued)

Date	Environmental protection policies, laws, regulations, etc.	Issuing authority
2009.8.26	*Measures for the Administration of the Special Fund for Substituting Subsidies with Rewards on Construction of Pipe Networks Supporting Urban Sewage Treatment Facilities*	The Ministry of Finance
2016.12.13	*Notice on Reducing the Vehicle Purchase Tax on Passenger Vehicles with an Engine Displacement of 1.6L or Less*	The Ministry of Finance, and the State Taxation Administration
2010.1.4	*Notice of the Ministry of Finance and the Ministry of Commerce on the Approval of Simultaneously Enjoying the Subsidy for Exchanging Second-hand Vehicles for New Ones and the Vehicle Purchase Tax Reduction Policies*	The Ministry of Finance, and the Ministry of Commerce
2010.4.19	*Notice of Opinions on Supporting Financing and Investment Policies and Measures for the Development of Circular Economy*	The NDRC, the People's Bank of China, the CBRC, and the China Securities Regulatory Commission (CSRC)
2010.5.21	*Credit Guidelines for Supporting Energy Conservation and Emission Reduction*	The People's Bank of China
2011.4.6	*Interim Measures for the Administration of the Subsidy Fund for Green Energy Demonstration County Construction*	The Ministry of Finance, the National Energy Administration, and the Ministry of Agriculture of the People's Republic of China (Ministry of Agriculture)
2011.10.29	*Notice on Carrying Out Trials for Trading Carbon Emissions Rights*	The General Office of NDRC
2012.1.5	*Notice on Issues concerning Preferential Policies on Enterprise Income Tax for Public Infrastructure Projects and Projects of Environmental Protection, Energy Saving and Water Conservation*	The Ministry of Finance, and the State Taxation Administration
2013.1.21	*Guiding Opinions on Carrying Out Compulsory Liability Insurance Trials for Environmental Pollution*	The Ministry of Environmental Protection, and the CIRC
2012.3.6	*Notice on Vehicle and Vessel Tax Policies for Energy-Efficient and New Energy Vehicles and Vessels*	The Ministry of Finance, the State Taxation Administration, the Ministry of Industry and Information Technology of the People's Republic of China (MIIT)
2012.8.20	*Provisions on the Collection and Management of Funds for the Disposal of Waste Electrical and Electronic Products*	The State Taxation Administration

(continued)

Table 3.11 (continued)

Date	Environmental protection policies, laws, regulations, etc.	Issuing authority
2013.1.14	*Notice on Further Implementing Air Quality Monitoring and Early Warning under Heavy Air Pollution*	The General Office of the Ministry of Environmental Protection
2013.2.27	*Announcement on Implementing Special Emission Limit of Air Pollutants*	The Ministry of Environmental Protection
2013.7.4	*Implementation Plan for the Pilot Program on the Replacement of Old Products with Remanufactured Products*	The NDRC, the Ministry of Finance, the MIIT, the Ministry of Commerce, and the GAQSIQ
2013.9.10	*Air Pollution Prevention and Control Action Plan*	The State Council
2013.12.2	*Notice on Improving Policies concerning the Fund for the Disposal of Waste Electrical Appliances and Electronic Products*	The Ministry of Finance, the Ministry of Environmental Protection, the NDRC, and the MIIT
2013.12.18	*Measures for the Evaluation of Enterprise Environmental Trustworthiness (for Trial Implementation)*	The Ministry of Environmental Protection, the NDRC, the People's Bank of China, and the CBRC
2014.1.14	*Interim Measures for the Administration of Water Pollution Prevention and Control Projects in Key Drainage Basins*	The General Office of NDRC
2014.6.27	*Key Performance Indicators of Green Credit Implementation*	The CBRC
2014.11.12	*Notice of the General Office of the State Council on Strengthening Environmental Supervision and Law Enforcement*	The General Office of the State Council
2014.8.6	*Guiding Opinions of the General Office of the State Council on Further Advancing the Pilot Work for the Paid Use and Trading of Emission Rights*	The General Office of the State Council
2014.10.20	*Notice on Environmental Protection Inspection Work System for Reform, Restructuring and Listing*	The Ministry of Environmental Protection
2014.12.2	*Guiding Opinions of the National Development and Reform Commission on Carrying out Public–Private-Partnership*	The NDRC
2014.12.22	*Guidelines for Enterprises Green Purchasing (for Trial Implementation)*	The Ministry of Commerce, the Ministry of Environmental Protection, and the MIIT
2014.12.27	*Opinions of the General Office of the State Council on Promotion of Third-Party Treatment of Environmental Pollution*	The General Office of the State Council

(continued)

Table 3.11 (continued)

Date	Environmental protection policies, laws, regulations, etc.	Issuing authority
2014.12.31	*Measures for the Administration of the Collection and Use of Sewage Treatment Fees*	The Ministry of Finance, the NDRC, and the MOHURD
2015.3.17	*Implementation Opinions on Encouraging and Directing the Participation of Private Investors in the Construction and Operation of Major Water Conservancy Projects*	The NDRC, the Ministry of Finance, and the Ministry of Water Resources of the People's Republic of China (Ministry of Water Resources)
2015.4.25	*Opinions of the CPC Central Committee and the State Council on Accelerating the Ecological Civilization Construction*	The CPC Central Committee, and the State Council
2015.6.18	*Measures for the Pilot Project of Collecting Volatile Organic Compounds Pollution Discharge Fee*	The Ministry of Finance, the NDRC, and the Ministry of Environmental Protection
2015.6.25	*Implementation Plan for the Environmental Protection "Frontrunner" System*	The Ministry of Finance, the NDRC, the MIIT, and the Ministry of Environmental Protection
2015.8.18	*Notice on Strengthening the Management of the Special Funds for Air Pollution Prevention and Control and Improving the Fund Usage Efficiency*	The Ministry of Finance, and the Ministry of Environmental Protection
2015.11.27	*Guiding Opinions on Strengthening the Construction of the Corporate Environmental Credit System*	The Ministry of Environmental Protection, and the NDRC
2015.12.31	*Guidelines for Issuance of Green Bonds*	The General Office of NDRC
2015.12.31	*Guiding Opinions on Promotion of Third-Party Treatment of Environmental Pollution in Coal-Fired Power Plants*	The NDRC, the Ministry of Environmental Protection, and the National Energy Administration
2016.1.11	*Notice on the Incentive Policy for New Energy Vehicle Charging Infrastructure and Strengthening the Promotion and Application of New Energy Vehicles in the "13th Five-Year Plan" Period*	The Ministry of Finance, the Ministry of Science and Technology of the People´s Republic of China (Ministry of Science and Technology), the MIIT, the NDRC, and the National Energy Administration
2016.1.11	*Notice of General Office of the National Development and Reform Commission on Effectively Performing the Key Tasks of Launching the National Carbon Emission Right Trading Market*	The General Office of NDRC
2016.2.17	*Guiding Opinions on Promoting Green Consumption*	The NDRC, and the Ministry of Environmental Protection

(continued)

Table 3.11 (continued)

Date	Environmental protection policies, laws, regulations, etc.	Issuing authority
2016.3.24	*Measures for the Administration of Guaranteed Buyout of Electricity Generated by Renewable Energy Resources*	The NDRC
2016.4.28	*Opinions of the General Office of the State Council on Improving the Compensation Mechanism for Ecological Protection*	The General Office of the State Council
2016.5.9	*Interim Measures for the Pilot Reform of Water Resource Tax*	The Ministry of Finance, the State Taxation Administration, and the Ministry of Water Resources
2016.5.27	*Notice of the National Development and Reform Commission and the National Energy Administration on Effectively Performing the Management on Guaranteed Buyout of Wind Power and Photovoltaic Power*	The NDRC, and the National Energy Administration
2016.10.31	*Notice on Issuing the "Interim Measures for the Administration of the Water Pollution Control and Treatment Project"*	The Ministry of Environmental Protection, and the Ministry of Housing and Urban–Rural Development
2016.11.24	*13th Five-Year Plan for Ecological Environmental Protection*	The State Council
2016.12.25	*Environmental Protection Tax Law of the People's Republic of China*	The NPC Standing Committee, and the Order of the President of the People's Republic of China
2017.3.18	*Notice of the General Office of the State Council on Forwarding the Implementation Plan of the Solid Waste Segregation System Made by the National Development and Reform Commission and the Ministry of Housing and Urban–Rural Development*	The General Office of the State Council
2008.7.1	*Provisions on the Administration of Environmental Impact Assessment of Marine Engineering*	The State Oceanic Administration
2017.12.17	*Plan for the Reform of the Compensation System for Damages to Ecology and Environment*	The General Office of the CPC Central Committee, and the General Office of the State Council
2018.1.30	*Implementation Plan for Central Treasury Incentive Polices on Promoting Ecological Conservation and Restoration of Yangtze River Economic Belt*	The Ministry of Finance, etc.
2020.4.23	*Notice of Improving the Policies on Government Subsidies for Promotion and Application of New Energy Vehicles*	The Ministry of Finance, etc.

(continued)

Table 3.11 (continued)

Date	Environmental protection policies, laws, regulations, etc.	Issuing authority
2018.6.16	*Opinions of the CPC Central Committee and the State Council on Comprehensively Strengthening Ecological and Environmental Protection and Resolutely Fighting the Uphill Battle for the Prevention and Control of Pollution*	The CPC Central Committee, and the State Council
2018.6.21	Opinions of the *National Development and Reform Commission on Innovating on Improving and Promoting the Green Development Price Mechanism*	The NDRC
2018.6.27	*Three-Year Action Plan for Defending the Blue Skies*	The State Council
2018.11.30	*Action Plan for Comprehensive Treatment of the Bohai Sea*	The Ministry of Ecology and Environment, the NDRC, the Ministry of Natural Resources, etc.
2018.12.28	*Action Plan for Establishing a Market-Oriented and Diversified Ecological Compensation System*	The NDRC, the Ministry of Finance, the Ministry of Natural Resources, etc.
2018.12.30	*Action Plan for Diesel Truck Pollution Control*	The Ministry of Ecology and Environment, the NDRC, the National Energy Administration, etc.
2019.4.26	*Notice of the Ministry of Housing and Urban–Rural Development, et al. on Full-Scale Launch of Household Waste Sorting in Cities at and above the Prefecture Level*	The MOHURD, etc.
2019.7.23	*Plan for the Natural Forest Protection and Restoration System*	The General Office of the CPC Central Committee, and the General Office of the State Council
2020.3.3	*Guiding Opinions on Building a Modern Environmental Governance System*	The General Office of the CPC Central Committee, and the General Office of the State Council

Data source Obtained upon sorting of environmental economic policies published on the website of the former Ministry of Environmental Protection and online platforms

Pollution issued by the CPC Central Committee and the State Council in 2018 further explicitly requires fighting three defense battles (blue skies, clear waters, and clean lands) and seven symbolic battles (blue skies, diesel truck pollution control, water source conservation, black and odorous water treatment, protection and restoration of the Yangtze River, comprehensive treatment of the Bohai Sea, and agricultural and rural pollution control), which shows the government's determination to curb environmental pollution.

Anatomizing the above policies, laws and regulations, we found that environmental protection policies could be classified into two types: administrative intervention (which means that environmental public goods are mainly supplied by the government) and market regulation (environmental public goods supplied by the market). Administrative intervention means that the government sector promulgates laws and regulations, e.g. developing unified technical standards and environmental risk levels, requiring related enterprises to obey, or punishing them pursuant to predetermined rules if they fail to obey. Market regulation depends heavily on the price mechanism, where the market decides everything and influences enterprise behavior by means of bank credit, fund supply and market prices, with the ultimate goal of curbing pollution and improving the human living environment. Furthermore, the 68 environmental protection policies, laws and regulations of China issued over the past decade could be sorted out. 60% of them belong to administrative intervention while market intervention policies only account for 40%. Certainly, administrative intervention has its inbuilt advantages, e.g. developing unified industry rules and punishment measures, which can greatly increase the corporate cost for violation of law by national legislation, and define the specific responsibilities for environmental protection to be assumed by the government and enterprises. Nevertheless, enterprises aim to maximize profits. It is impossible to enable enterprises to fully realize the importance of environmental protection and even actively participate therein in reliance upon administrative intervention alone. It should be particularly noted that some enterprises may conceal the facts by devising countermeasures against government policies. At the current stage, China's environmental protection only has few fund sources, so environmental pollution control often has a lack of financial support. In this case, if the government sector continuously applies market regulation policies, and enhance public participation and social supervision, enterprises will definitely attach greater importance to environmental protection.

Thus, the simultaneous use of administrative intervention and market regulation policies will get double results with half the effort in terms of the supply of environmental public goods.

3.3.2 Single Supply Mode of Environmental Public Goods

The supply of public goods has always been a topic discussed in the academic circles. Because of the "free-rider" problem, more economists suggested the government supply of public goods, including environmental public goods.

As regards environmental governance, the Chinese government tends to impose taxes on polluters, i.e. forcing enterprises to pay the consequences. The government tries to improve the negative externality of society by imposing taxes. For the industrially discharged "three wastes" pollution sources, the government did not put identical efforts into waste gas, wastewater and solid waste control (see Fig. 3.15). Since 2013, the government had significantly increased its efforts to curb waste gas, from the investment of RMB 20 billion in the past few years to RMB 64.091 billion in 2013 (only RMB 27.526 billion in 2007), which increased to the highest investment in waste gas control—RMB 78.939 billion in 2014, equivalent to 186.78% of the investment in 2007. In 2017, the government reduced its investment in waste gas, wastewater and solid waste control, completed an investment of RMB 44.626 in waster gas control projects, a decrease of 20.52%; completed an investment of RMB 7.638 billion in waste water control projects, a decrease of 29.44%; completed an investment of RMB 1.274 billion in solid waste control projects, a decrease of 72.7%. However, it greatly increased its investment in curbing noise and other pollution, completed an investment of RMB 129 million in noise control projects, an increase of 106.25%; completed an investment of RMB 14.487 billion in other projects, an increase of 42.03%. Though the government invested a large quantity of financial and material resources in curbing waste gas, the reality did not match the government's investment.

The government supply of public goods is likely to cause government failure. To put it more specifically: (1) fiscal decentralization and the competition among local governments adversely affect the investment structure of local public goods, and the existing financial system arrangements separate the domestic market, resulting in under-investment in public goods actually needed; (2) the government performance assessment mechanism focused on quantity expansion causes local officials to heavily invest in projects which are capable of producing significant direct economic benefits in a short term, with long-lasting undersupply of intergenerational (environmental) public goods.

China has successfully issued a large number of policies to promote the supply mode diversification of environmental public goods. The implementation of such policies as the pilot emission right trading and the green bonds officially issued in 2015 has effectively eased the increasingly serious environmental pollution to some extent. However, the above policies are merely the source diversification of government supply funds, but not the supply mode diversification.

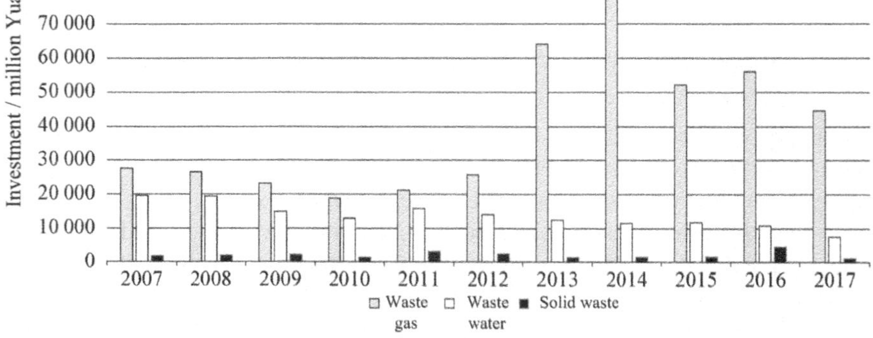

Fig. 3.15 Completed investment in China's industrial pollution control. *Data source* Obtained from the data of the former Ministry of Environmental Protection

At present, the supply modes of public goods comprise government supply, market supply, voluntary supply and mixed supply, i.e. the PPP supply mode [127]. The "free-rider" problem of public goods can be extended to the conclusions that government governance is the most efficient, and the government may fail; then, the market mechanism can be introduced into the supply of public goods, followed by a market-oriented and socialized multi-subject governance mode. Which operation mechanism can reveal residents' preference for public services to the maximum extent? Which mechanism is more effective for China's environmental pollution control? These are our major concerns at the present time.

The single supply mode of environmental public goods is definitely prejudicial to pollution control. Enterprises are an important source of pollution, but with the in-depth research and data analysis on residents' behavior, we found that residents as the demand side of clean environment have increasingly become the supply side of pollution. Though the enterprise pollution source is now declining, household wastewater and waste pollution caused by residents are becoming increasingly severe. If it is possible to enhance residents' environmental awareness, change their behavior pattern, and mobilize residents, enterprises and other social groups to actively cooperate with the government in forming a compound supply mode of environmental public goods, our living environment can be improved.

3.3.3 Inadequate Policy Efforts in Internalizing the Externality Problem of Polluters

As regards the settlement of the externality problem, government regulation and market regulation are frequently mentioned in economic research, i.e. the use of Pigovian taxes and the Coase theorem for settlement. The biggest advantage of government regulation lies in its authority and rapidity in issuing orders. However, it has a deficiency in efficiency, i.e. government failure.

According to the Coase theorem, as long as property rights are clearly defined, and transaction costs are zero or tiny, the final market outcome is always efficient regardless of whichever is initially awarded the property rights [128]. To effectuate this theorem, two conditions must be satisfied: clearly defined property rights, and tiny or even zero transaction costs. It also means that the polluting party and the polluted party should coordinate with each other for settlement on a voluntary basis. In the case that the above conditions are satisfied, the allocation of social resources can be optimized upon the entry into effect of the market mechanism. The Coase theorem belongs to an idealized solution in reality, and its required conditions are basically impossible to meet. Thus, market failure is inevitable.

In reality, our pollution control policies are basically dominated by administrative intervention (e.g. closing, suspension, merging and shifting) and tax increase. To intensify environmental governance and reduce pollutant discharge, China has established a system of fee charging for pollutant discharge since 1979, i.e. taking the economic means of fee collection to push enterprises into supplementing environmental governance funds, which has played a significant role in pollution prevention and control and environmental protection. However, there are some problems in the execution of the system, which affects the normal functioning thereof. Moreover, there is a lack of incentive to stimulate polluters' internalization behavior of reducing pollution by improving technology, which causes enterprises to lag behind in R&D inputs and technical innovation, and even leads to the "hide-and-seek" between polluters and inspection authorities: when superior units send some personnel to inspect pollution status, polluters may suspend or reduce production, or even choose not to discharge pollutants in daytime, but to secretly discharge wastewater and waste residue in the dead of night.

In fact, there is a strategy for internalizing the externality problem, i.e. "holding those who cause pollution responsible for eliminating such pollution". For instance, polluters take the initiative in cooperating with sewage treatment plants, or replace desulphurization equipment. We suggest the government stimulate the subject behavior with positive externality, encourage enterprises to internalize the externality problem. If the government can take effective measures to motivate polluters to internalize negative externality, it is possible to not only fill the loophole of "devising countermeasures against government policies", but also solve the shortage of pollution control funds and the technological backwardness of enterprises.

3.3.4 A Lack of Effective Measures to Motivate Residents to Participate in Pollution Control

The government, enterprises and residents are all participants in the operation process of the market economy system. So far, the participants in environmental governance—the government, the market and the public (dominated by volunteer organizations at present)—all have failures. We cannot effectively solve environmental pollution in reliance upon any of them alone.

At present, as the demand side of clean environment, residents have not fully participated in pollution control, or even have been basically absent from it (only a small minority of them have made innovations, participated in tree planting or other volunteer activities through volunteer organizations). However, with the most pressing need for environmental cleanliness, residents stay aloof. This will surely affect the performance of environmental governance. If residents can be motivated to participate in the supply of environmental public goods, double results will be achieved with half the effort. On the one hand, residents are one of the pollution sources. After participating in environmental governance, they will be aware of the environmental damages arising from their behavior, and will definitely restrain such behavior to some extent, e.g. changing their travel mode, saving water resources, and conducting waste sorting. On the other hand, when residents participate in environmental governance, the government can motivate them to pay for environmental protection by some feasible measures, e.g. setting an appropriate discount rate to encourage young and middle-aged residents to actively pay for environmental protection, and giving them discounts after they get old. In this way, the current financial strain of environmental governance can be relieved. In addition, after participating in environmental governance, residents can mobilize their relatives and friends to take joint action by their words and deeds, and especially influence their descendants' behavior to make the entire people have a sense of urgency, which will certainly have a profound impact on the environmental aspiration inherited by generation after generation. The public participation in environmental governance is more conducive to social supervision, which can serve as a supplement to legal loopholes to strengthen the government responsiveness, and enable enterprises and the public to restrain themselves to some extent and to march towards the common vision—blue skies and white clouds.

Certainly, to build a multi-subject environmental governance mechanism covering residents to enable these subjects to complement and coordinate mutually and to push China's environmental public goods to a new height, there must be a precondition—residents' participation. Therefore, how to motivate residents to participate in the supply of environmental public goods and change the single supply mode becomes a focus of the discussion herein.

Chapter 4
Theoretical Modeling for Residents' Participation in Supply of Environmental Public Goods

Based on a model framework with private (or family) pursuit of utility maximization, this chapter interprets residents' motivation for voluntary supply of environmental public goods on a micro level. Through research, we intend to demonstrate that residents have the willingness to supply environmental public goods, interpret the phenomenon of private voluntary supply existing in reality from a theoretical perspective, and hope to draw a conclusion concerning the conditions of residents' voluntary supply in reliance upon the model, and provide some helpful ideas for the government to promote residents' voluntary supply.

4.1 Basic Model Assumption

We consider a two-period OLG model, and simplify the economy into the composition of two-period (the young and middle-age period t, and the old-age period $t + 1$) representative individuals, perfectly competitive firms and the government. The modeling is extended from the government supply of environmental public goods, and residents as the demand side of quality environment are brought into the model to theoretically demonstrate the benefits that residents can bring as the supply side.

4.1.1 Individual

We assume that the population growth rate is 0; the representative individuals in the period t put in 1 unit of labor, get the wage W_t, and choose the current consumption C_t, the savings S_t, the input in improving environmental quality M_t and the consumption in old age C_{t+1}. The consumption and the environmental quality in the two periods directly influence individual utility—U (C_t, C_{t+1}, E_t, E_{t+1}), in which E_t and E_{t+1}

J. Yang, *Effective Supply Mechanism and Path of Environmental Public Goods in China*, Public Economy and Urban Governance in China,
https://doi.org/10.1007/978-981-99-2939-9_4

respectively signify the environmental quality in the periods t and $t+1$. Individual goal is to obtain utility maximization through the above decisions. The utility function is expressed by Eq. (4.1).[1]

$$U(C_t, C_{t+1}, E_t, E_{t+1}) = \ln C_t + \phi_t \ln E_t + \frac{1}{1+\rho}(\ln C_{t+1} + \phi_{t+1} \ln E_{t+1}) \quad (4.1)$$

where ϕ_t signifies the degree to which consumers in the period t attach importance to environmental quality; ϕ_{t+1} signifies the degree to which consumers in the period $t+1$ attach importance to environmental quality; ρ is the discount rate. Due to the inconsistency between the utility brought by the current consumption and that brought by the same consumption in the future, in case of a bigger ρ, consumers will be more inclined towards current consumption, and reduce consumption in the next period.

With limited income W_t, individuals choose consumption, savings and the input in improving environmental quality M_t, so Eq. (4.2) is satisfied[2]:

$$\begin{aligned} C_t + S_t + M_t &= W_t + TR_t^d \\ TR_t^d &= \mu^d M_{t-1} \end{aligned} \quad (4.2)$$

where S_t signifies the savings in the period t; $TR^d{}_t$ is the government's transfer payment to individuals; μ^d is the medical or pension subsidy rate set for individual payment of environmental protection funds by the government or the inheritance tax discount rate in the period $t+1$.

The savings S_t in the period t are the source of consumption in the old-age period $t+1$. We assume that the nominal interest yield rate in the period $t+1$ is r_{t+1}, then the following equation can be obtained:

$$C_{t+1} = S_t(1 + r_{t+1}) \quad (4.3)$$

By reference to the practice of John and Pecchenino [12] and Ono [121], we express the environmental quality equation as[3]:

$$\begin{aligned} E_{t+1} &= (1+b)E_t - P_t - \beta(C_t + C_{t+1}) + \theta T_t + \varepsilon M_t, \ (\beta, \varepsilon, \theta > 0) \\ P_t &= \bar{\phi} K_t \end{aligned} \quad (4.4)$$

[1] In the utility functions presented by Ono (1996), David (2004) and Hong Shulin (2006), the environmental quality E_t in the current period was not considered. They considered that the current environmental quality was fixed, and mainly depended on the capital and pollution reduction in the previous period. However, this study holds that the current environmental quality not only depends on the previous period, but also relates to the production and consumption activities in the current period.

[2] The individual payment of environmental public goods is calculated according to the regional residents' willingness to pay, so $M_t = M_{t-1}$ in this book.

[3] Their papers only consider the environmental pollution arising from private consumption, but this study takes into account the environmental pollution caused by private consumption and enterprise production.

where b is the spontaneous change rate or self-purification rate of environmental quality, $b \in [-1, 0) \cup (0, 1]$; P_t is the pollution caused by the production activities of enterprises, showing the environmental quality deterioration arising from the capital participation in production activities; β is the degree of pollution effect resulting from consumption activities; ε signifies the effective coefficient of individual input in environmental quality control; εM_t is the improvement of environmental quality brought by individual input in improving environmental quality M_t for reducing pollution; θ signifies the effective coefficient of the government's environmental quality control; T_t is the environmental tax (Pigovian tax) imposed on enterprises by the government for improving environmental quality.

The family goal is to maximize their own utility, i.e. Equation (4.1), under the budget constraints of Eqs. (4.2) and (4.3).

4.1.2 *Firms*

Perfectly competitive firms

$$Y_t = AF(K_t, L_t) = AK_t^\alpha L_t^{1-\alpha}$$

where Y_t is the yield in the period t; A represents technological level; K_t is the capital input in the period t; L_t is the labor input in the period t.

Representative individuals are assumed to use 1 unit of labor every period, so the above equation can be simplified as follows:

$$Y_t = AF(K_t) = AK_t^\alpha \tag{4.5}$$

$$\pi = Y_t - r_t K_t - \delta K_t - W_t - \tau P_t \tag{4.6}$$

$$r_t = \alpha A K_t^{\alpha-1} - \delta - \tau\varphi \tag{4.7}$$

where α is the contribution rate of capital to output; r is the capital price; π is the profit; δ is the capital depreciation rate; φ is the proportion of pollution arising from capital input; τ is the pollution tax rate.

According to the full and net distribution by perfectly competitive firms in Euler's theorem, the surplus of output is distributed as wages, and then the labor wage is obtained:

$$W_t = (1 - \alpha)Y_t = (1 - \alpha)AK_t^\alpha \tag{4.8}$$

The commodity market clearing meets the condition that supply equals demand, actual gross product value = consumption + total investment = consumption + net investment + depreciation, i.e.:

$$Y_t = C_t + C_{t+1} + K_{t+1} - (1 - \delta)K_t \tag{4.9}$$

4.1.3 Government

Due to the market failure resulting from the double externalities of the environment, the government sector imposes the environmental pollution tax T_t and puts the environmental protection fund M_t paid by individuals of their own accord into environmental governance. M_t is the amount paid according to residents' marginal willingness, which is measured by the happiness measurement method.

$$T_t = \tau\phi K_t \tag{4.10}$$

On the one hand, the government puts the environmental tax revenue into the improvement and control of environmental quality, and stimulates enterprises which have performed well in pollution control by means of government subsidies, e.g. giving enterprises with technological progress a reward $TR^f{}_t$. On the other hand, for residents who are willing to pay the environmental protection fund in the period t (young and middle-age period), the government transfers $TR^d{}_t$ to individuals in the period $t + 1$ (old-age period) by means of increasing medical insurance, old-age insurance or inheritance tax deductions. Thus, the following equations are obtained:

$$T_t + M_t = G_t + TR_t^f + TR_t^d \tag{4.11}$$

$$TR_t^f = \mu^f \tau\varphi K_t, \quad TR_t^d = \mu^d M_t \tag{4.12}$$

where μ^f signifies the proportion of government rewards given to enterprises which have performed well in pollution control (technological progress); μ^d signifies the proportion of government incentives (subsidies in the old-age period) given to residents who have paid the environmental protection fund.

4.2 Basic Model of Environmental Public Goods Fully Provided by the Government

How is the market equilibrium situation when the government serves as a single subject of environmental governance? This part mainly probes into the Pigovian tax. We describe the government provision of environmental public goods as G-P model (Government Provide Model).

4.2.1 Analysis on the Model of Government Provision of Environmental Public Goods

Air pollution has the typical characteristics of forward intergenerational public goods, and thus has significant generational and intergenerational externalities. Guided by the thought of "free-riding", residents could only rely on the government to improve environmental quality by imposing a pollution tax on enterprises so as to rectify the external diseconomy on the market. In case $M_t = 0$ in Eqs. (4.2) and (4.4), then $T_t = \tau P_t = \tau\varphi K_t$; government policies are aimed at ensuring stable environmental quality in the periods t and $t + 1$, i.e. $E_t = E_{t+1}$.

The revenue budget constraint turns into $C_t + S_t = W_t$. Under the above constraint and Eq. (4.3), individuals choose C_t, C_{t+1} and S_t to maximize individual utility Eq. (4.1).

$$
\begin{aligned}
&\text{Max}\, U(C_t, C_{t+1}, E_t, E_{t+1})\\
&U(C_t, C_{t+1}, E_t, E_{t+1})\\
&\quad = \ln C_t + \varphi_t \ln E_t + \frac{1}{1+\rho}(\ln C_{t+1} + \varphi_{t+1} \ln E_{t+1})\\
&\quad = \ln C_t + \varphi_t \ln E_t + \frac{1}{1+\rho}\{\ln C_{t+1} + \varphi_{t+1} \ln[(1+b)E_t\\
&\qquad - P_t - \beta(C_t + C_{t+1}) + \theta T_t]\}
\end{aligned}
$$

Building the Lagrangian function and eliminating the shadow price, we can obtain:

$$
\frac{C_{t+1}}{C_t} = \frac{1 + r_{t+1}}{1+\rho} \tag{4.13}
$$

$$
E_t = \frac{\theta}{b}T_t - \frac{\varphi}{b}K_t - \frac{\beta}{b}(C_t + C_{t+1}) \tag{4.14}
$$

According to Eqs. (4.5)–(4.9) and (4.14), we can obtain:

$$
K_{t+1} = AK_t^{\alpha} + (1-\delta)K_t - \frac{b}{\beta}E_t - \frac{\varphi}{\beta}(1-\theta\tau)K_t \tag{4.15}
$$

When the economy reaches a steady state, $K_{t+1} = K_t$, then

$$E_t = \frac{\beta}{b}AK_t^{\alpha} - \frac{K_t}{b}(\beta\delta + \varphi - \theta\tau) \tag{4.16}$$

In steady state, to reach the optimum environmental quality, capital $\overline{K}$ is required to be:

$$\overline{K_t} = \left(\frac{\beta\delta + \varphi - \theta\tau}{\alpha\beta A}\right)^{\frac{1}{\alpha-1}} \tag{4.17}$$

In this case, environmental quality $\overline{E}_t = \overline{E}_{t+1}$

$$\overline{E}_t = \frac{\beta}{b}A\overline{K}_t^{\alpha} - \frac{\overline{K}_t}{b}(\beta\delta + \varphi - \theta\tau)$$

That is

$$\overline{E_t} = \frac{1}{b}\left(\frac{\beta\delta + \varphi - \theta\tau}{\alpha\beta A}\right)^{\frac{\alpha}{\alpha-1}}\left(\beta A - \frac{1}{\alpha\beta A}\right) \tag{4.18}$$

4.2.2 Conclusions on the Model of Government Provision of Environmental Public Goods

The above inference and Eq. (4.18) show that smaller b, φ, δ and β will lead to higher environmental quality E; bigger θ and τ will lead to higher environmental quality E. In other words, in steady state:

(1) In case $\mathrm{d}K/\mathrm{d}\theta > 0$, $\mathrm{d}E/\mathrm{d}\theta > 0$, higher pollution control efficiency is beneficial to the government's accumulation of higher capital, and brings higher environmental quality and economic growth;
(2) In case $\mathrm{d}K/\mathrm{d}\tau > 0$, $\mathrm{d}E/\mathrm{d}\tau > 0$, imposing a higher environmental tax on enterprises is more beneficial to capital accumulation and environmental quality improvement;
(3) In case $\mathrm{d}\overline{K}/\mathrm{d}\varphi < 0$, $\mathrm{d}\overline{E}/\mathrm{d}\varphi < 0$; $\mathrm{d}\overline{K}/\mathrm{d}\beta < 0$, $\mathrm{d}\overline{E}/\mathrm{d}\beta < 0$, a higher degree of pollution caused by enterprises (bigger pollutant discharge coefficient) will lead to heavier pollution from residents' consumption activities, slower capital accumulation and worse environmental quality;
(4) In case $\mathrm{d}\overline{K}/\mathrm{d}b < 0$, $\mathrm{d}\overline{E}/\mathrm{d}b < 0$, a higher rate of environmental degradation is more disadvantageous to capital accumulation and environmental improvement;

(5) In case $d\overline{K}/d\delta < 0$, $d\overline{E}/d\delta < 0$, a higher depreciation rate of enterprise capital will lead to the decline of capital accumulation and the deterioration of environmental quality.

Furthermore, with more advanced technology in economy and bigger capital output elasticity α, more capital will be accumulated, and environmental quality will be further improved.

4.3 Model of Government-Led Provision of Environmental Public Goods by the Government, Residents and Enterprises

We describe the joint provision of environmental public goods by residents, enterprises and the government as REG-P model (Residents, Enterprises and Government Provide Model).

To reduce residents' "free-riding" behavior and facilitate the sustainability of individual consumption, the government encourages residents to actively participate in environmental governance, advocates low-carbon consumption, give rewards to those individuals who are willing to pay the pollution control fee by means of increasing medical insurance, old-age insurance or inheritance tax deductions in their old-age period, and still imposes a carbon tax on polluters (pollutant discharge enterprises) for the purpose of remedying market failures.

4.3.1 Modeling for Joint Provision of Environmental Public Goods

In case $M_t \neq 0$ in the aforesaid Eqs. (4.2) and (4.4), government policies are still aimed at ensuring stable environmental quality in the periods t and $t+1$, i.e. $E_t = E_{t+1}$.

The revenue budget constraint turns into $C_t + S_t + M_t = W_t + TR_t^d$. Under the constraint and Eq. (4.3), individuals choose C_t, M_t and S_t to maximize individual utility Eq. (4.1).

$$
\begin{aligned}
&\mathrm{Max}U(C_t, C_{t+1}, E_t, E_{t+1}) \\
&U(C_t, C_{t+1}, E_t, E_{t+1}) \\
&\quad = \ln C_t + \varphi_t \ln E_t + \frac{1}{1+\rho}(\ln C_{t+1} + \varphi_{t+1} \ln E_{t+1}) \\
&\quad = \ln C_t + \varphi_t \ln E_t + \frac{1}{1+\rho}\{\ln C_{t+1} + \varphi_{t+1} \ln[(1+b)E_t \\
&\qquad - P_t - \beta(C_t + C_{t+1}) + \theta T_t + \varepsilon M_t]\}
\end{aligned}
$$

We build the Lagrangian function

$$L = \ln C_t + \varphi_t \ln E_t + \frac{1}{1+\rho}\{\ln C_{t+1} + \varphi_{t+1} \ln[(1+b)E_t - P_t - \beta(C_t + C_{t+1}) + \theta T_t + \varepsilon M_t]\} + \lambda\left(W_t + TR_t^d - C_t - S_t - M_t\right)$$

where, λ denotes the shadow price.

According to residents' choices C_t, S_t and M_t, i.e. $\partial L/\partial C_t$, $\partial L/\partial C_{t_0}$ and $\partial L/\partial M_{t_s}$, the shadow price is eliminated.

$$\frac{\partial L}{\partial C_t} = \frac{1}{C_t} - \frac{\beta \cdot \phi_{t+1}}{(1+\rho)E_{t+1}} - \lambda = 0$$

$$\frac{\partial L}{\partial S_t} = \frac{(1+r_{t+1})}{(1+\rho)C_{t+1}} - \frac{\beta \cdot \phi_{t+1}(1+r)}{(1+\rho)E_{t+1}} - \lambda = 0$$

$$\frac{\partial L}{\partial M_t} = \frac{\varepsilon \cdot \phi_{t+1}}{(1+\rho)E_{t+1}} + \lambda\mu^d - \lambda = 0$$

We can obtain:

$$E_{t+1} = \frac{1}{(1+\rho)}C_t + \frac{\varepsilon \cdot r \cdot \phi_{t+1}}{(1+\rho)(1+r)(1-\mu^d)}C_t \tag{4.19}$$

$$C_t = \frac{(1+\rho)(1+r)(1-\mu^d)}{(1+r)(1-\mu^d) + \varepsilon \cdot r \cdot \phi_{t+1}}E_{t+1} \tag{4.20}$$

$$C_{t+1} = \frac{(1+r)(1-\mu^d)}{\beta \cdot \phi_{t+1}(1+r)(1-\mu^d) + \varepsilon \cdot \phi_{t+1}}E_{t+1} \tag{4.21}$$

According to Eqs. (4.5)–(4.9), the enterprise profit is maximized:

$$\pi = \alpha A K_t^\alpha - (r + \delta + \tau \cdot \varphi)K_t$$

$$\frac{\partial \pi}{\partial K_t} = 0$$

$$K_t = \left(\frac{r+\delta+\tau\cdot\varphi}{\alpha^2 \cdot A}\right)^{\frac{1}{\alpha-1}}$$

$$K_{t+1} = AK_t^\alpha + (1-\delta)K_{\bar{t}} - \left[\frac{(1+\rho)(1+r)\left(1-\mu^d\right)}{(1+r)\left(1-\mu^d\right) + \varepsilon \cdot r \cdot \phi_{t+1}} + \frac{(1+r)\left(1-\mu^d\right)}{\beta \cdot \phi_{t+1}(1+r)\left(1-\mu^d\right) + \varepsilon \cdot \phi_{t+1}}\right]E_{t+1} \tag{4.22}$$

When the economy reaches a steady state, $K_{t+1} = K_t$, $E_{t+1} = E_t$, then

$$E_t = \frac{AK_t^{\alpha} - \delta K_t}{\left[\frac{(1+\rho)(1+r)(1-\mu^d)}{(1+r)(1-\mu^d)+\varepsilon \cdot r \cdot \phi_{t+1}} + \frac{(1+r)(1-\mu^d)}{\beta \cdot \phi_{t+1}(1+r)(1-\mu^d)+\varepsilon \cdot \phi_{t+1}}\right]}$$

$$E_t = \frac{A\left(\frac{r+\delta+\tau \cdot \varphi}{\alpha^2 \cdot A}\right)^{\frac{\alpha}{\alpha-1}} - \delta\left(\frac{r+\delta+\tau \cdot \varphi}{\alpha^2 \cdot A}\right)^{\frac{\alpha}{\alpha-1}}}{\left[\frac{(1+\rho)(1+r)(1-\mu^d)}{(1+r)(1-\mu^d)+\varepsilon \cdot r \cdot \phi_{t+1}} + \frac{(1+r)(1-\mu^d)}{\beta \cdot \phi_{t+1}(1+r)(1-\mu^d)+\varepsilon \cdot \phi_{t+1}}\right]} \tag{4.23}$$

and

$$C = \frac{(1+\rho)(1+r)(1-\mu^d)}{(1+r)(1-\mu^d)+\varepsilon \cdot r \cdot \phi_{t+1}}$$

$$D = \frac{(1+r)(1-\mu^d)}{\beta \cdot \phi_{t+1}(1+r)(1-\mu^d)+\varepsilon \cdot \phi_{t+1}}$$

As

$$\frac{\partial C}{\partial \mu^d} > 0, \quad \frac{\partial D}{\partial \mu^d} < 0$$

and $AK^{\alpha} > \delta K$

Then

$$\frac{dE_{t+1}}{d\mu^d} = \frac{-(AK^{\alpha} - \delta K)(C' + D')}{(C + D)^2} > 0 \tag{4.24}$$

$$K_t = \frac{1}{\delta}\left\{A\left(\frac{r+\delta+\tau \cdot \varphi}{\alpha^2 \cdot A}\right)^{\frac{\alpha}{\alpha-1}} - (C + D)E_t\right\} \tag{4.25}$$

Similarly,

$$\frac{\partial K_t}{\partial \mu^d} > 0 \tag{4.26}$$

4.3.2 Conclusions on the Model of Joint Provision of Environmental Public Goods

Equations (4.24) and (4.26) suggest that when residents pay the pollution control fee according to their marginal willingness to pay, they will enjoy higher old-age subsidies or medical subsidy and better environmental quality in the old-age period $(t + 1)$, and promote economic growth. Conversely, it can be understood that when the government gives higher subsidies in the period $t + 1$ to those residents who have paid pollution control fee, it can better facilitate their payment of pollution control fee in the young period (period t). Improving environmental quality is beneficial to not only their physical health, but also that of their descendants. Such action can ensure sustainable economic and environmental growth.

4.4 Game Analysis on Supply of Environmental Public Goods

Through the above theoretical analysis, we have understood that the joint provision of environmental public goods by the government, residents and enterprises could bring multiple dividends. At present, we need to probe into the following issues: How should we motivate enterprises to curb pollutant discharge voluntarily? Why could individuals voluntarily provide environmental public goods? Through building a game model for voluntary pollution control by enterprises and governmental punishments for pollutant discharge behavior, and analyzing the model of "Boxed Pigs Game" between the government sector and the private sector during the government provision of environmental public goods, we can obtain a feasible way for multi-subject supply of environmental public goods.

4.4.1 Mechanism for Voluntary Provision of Environmental Public Goods by Enterprises

According to the results obtained from the above data analysis, we can probe into the building of a game model between the government and enterprises. Suppose that enterprises are willing to curb pollutant discharge, i.e. willing to provide environmental public goods for pollutant discharge. Their required cost is denoted by C; profit obtained in reliance upon their business is denoted by w; required compensation for the loss to surrounding residents arising from a failure to curb their pollutant discharge is denoted by ε; their generated social benefit is denoted by α; the social loss resulting from their pollutant discharge is denoted by β. In case enterprises are unwilling to provide environmental public goods, i.e. enterprises bear the cost for all environmental public goods, the government may choose to supervise enterprises, and impose a penalty on them with the penalty amount γ when the pollution caused by them severely affects the environment. Besides, to grasp the specific pollution situation of enterprises, the government needs to supervise them, with the supervision cost s.

Meanwhile, this book assumes that the following two conditions are satisfied: first, in case enterprises choose not to curb pollutant discharge, the sum of their compensation for surrounding residents and received penalty is greater than the treatment cost when they choose to curb pollution; second, in case enterprises do not provide environmental public goods, their received penalty is greater than the

Table. 4.1 Game matrix between the government and enterprises

Matrix		Government	
		Penalty	No penalty
Enterprises	Pollution control	$\underline{\omega - C}, \alpha - s$	$\omega - C, \underline{\alpha}$
	No pollution control	$\omega - \varepsilon - \gamma, \underline{\alpha + \gamma - s - \beta}$	$\underline{\omega}, \alpha - \beta$

loss caused to society when they do not provide such goods. Based on the above assumption, the game matrix between them is shown in Table 4.1.

Obviously, the above game matrix suggests that no pure-strategy Nash equilibrium exists in the game process. Starting from the definition, we now seek a mixed strategy Nash equilibrium.

Suppose the probability of pollution control by enterprises is p, then the probability of no pollution control is $1 - p$.

Suppose the government's probability of pollutant discharge supervision on enterprises is q, then the probability of no supervision is $1 - q$. The revenue function of enterprises is:

$$U_{\text{enterprise}} = p[q(\omega - C) + (1 - q)(\omega - C)] + (1 - p)[q(\omega - \varepsilon - \gamma) + (1 - q)\omega] \tag{4.27}$$

The revenue function of the government is:

$$U_{\text{government}} = q[p(\alpha - s) + (1 - p)(\alpha + \gamma - s - \beta)] + (1 - q)[p\alpha + (1 - p)(\alpha - \beta)] \tag{4.28}$$

Calculating the partial derivative with respect to Eq. (4.17), we can obtain:

$$\frac{\partial U \text{ enterprise}}{\partial p} = q\varepsilon + q\gamma - C \tag{4.29}$$

Enterprises aim to maximize their own interests. Thus, setting Eq. (4.29) = 0, we can obtain:

$$q = \frac{C}{\varepsilon + \gamma} \tag{4.30}$$

Calculating the partial derivative with respect to Eq. (4.28), we can obtain:

$$\frac{\partial U \text{ government}}{\partial p} = \gamma - s - p\gamma \tag{4.31}$$

The government aims to maximize its own interests. Thus, setting Eq. (4.31) = 0, we can obtain:

$$p = \frac{\gamma + s}{\gamma} = 1 + \frac{s}{\gamma} \tag{4.32}$$

Based on the above calculation results, we analyze the provision of environmental public goods by enterprises.

First, we analyze p. According to Eq. (4.32), the value of p—probability of pollution control by enterprises which are willing to provide environmental public goods, is negatively correlated with their received penalty γ, but positively correlated with the government's supervision cost s. In other words, in case enterprises receive a greater penalty for failing to curb the pollution arising in their production process, they are more likely to voluntarily provide environmental public goods so as to reduce pollutant discharge. Meanwhile, in case the government has a greater difficulty in supervising the pollutant discharge by enterprises, i.e. when the government supervises enterprises at an excessive cost, enterprises will have a smaller probability to voluntarily provide environmental public goods, because when their pollutant discharge behavior is less likely to be identified, they are more willing to risk discharging pollutants.

Then, we analyze q. According to Eq. (4.30), the value of q—the government's probability to supervise enterprises and punish them when their pollutant discharge exceeds the permitted limit, is negatively correlated with ε—their compensation for residents arising from a failure to curb their pollutant discharge, and their received penalty γ, but positively correlated with their required cost C for providing environmental public goods. In other words, in case enterprises choose not to voluntarily provide environmental public goods, when they have to make higher compensation to surrounding residents and pay a higher penalty, they will be more afraid to discharge pollutants, or as enterprises have to pay an excessive marginal cost for not curbing pollutant discharge, the government may reduce the probability of supervision accordingly. Furthermore, in case enterprises need to pay an excessive cost for providing environmental public goods, they tend not to provide such goods. This requires the government to focus on supervising the pollutant discharge by enterprises, and the value of q will be larger accordingly.

To sum up, the following conclusions can be obtained herein. On the one hand, to make more enterprises willing to provide environmental public goods and control pollution from the source, powerful supervision and severe punishments can be implemented to increase the p value of enterprises, and to spur them to choose to provide environmental public goods. On the other hand, the government may give certain subsidies to enterprises which voluntarily provide environmental public goods so as to reduce their probability of discharging pollutants but not curbing pollution.

There is still much to be improved in the *Environmental Protection Law* of China, For example, as regards the regulations on the discharge of pollutants by enterprises, most of them require enterprises to pay a percentage of pollutant discharge fee; for those enterprises discharging pollutants in excess of prescribed standards, related government regulators will require them to pay a certain amount of penalty. However, the existing laws on environmental pollution prevention and control set a

Table. 4.2 Game payoff matrix between the government and the private sector

Matrix		Government	
		Provide	Not provide
Private sector	Provide	$U_g - C_g, U_p - (C - C_g)$	$U_g, U_p - c$
	Not provide	$U_g - c, U_p$	0, 0

low standard for pollutant discharge fee, with weak punishment, and a large number of regulations and laws only prescribe a low payable amount for pollution behavior. Some materials suggest that the fee for excessive pollutant discharge paid by polluters is only equivalent to 10–15% of the pollution control fee, far below than the normal pollution control fee.[4] In addition to collecting pollutant discharge fees and imposing a penalty on enterprises causing severe pollution, the government should consider working with enterprises to curb pollution. For many enterprises, their voluntary pollution control requires not only such reverse incentives as penalties and pollutant discharge fees, but also positive incentives like government investment. According to the above analysis, when enterprises can obtain subsidies to their satisfaction, namely when their marginal cost for voluntary pollution control is lower than their marginal revenue, enterprises will choose to provide environmental public goods. With a strong legal guarantee, and explicit provisions on the subsidies for different enterprises and businesses and the punishment standards for enterprises at different pollutant discharge levels, the phenomena like power rent-seeking could be reduced.

4.4.2 "Boxed Pigs Game" Between the Government Sector and the Private Sector

In the supply process of environmental public goods, there are some similarities to the "Boxed Pigs Game". We assume that the provision cost of environmental public goods needed in society is C. In case the government and the private sector jointly provide environmental pubic goods, the portion provided by the government is C_g, and then the portion provided by the private sector is $C - C_g$. After the supply of environmental public goods, the government's revenue is U_g, and the revenue of the private sector is U_p. Meanwhile, we also assume that when environmental public goods are only provided by the private sector, the revenue thereof U_p is smaller than the cost required for providing such goods. According to the above assumptions, the game payoff matrix between them can be obtained, as shown in Table 4.2.

Ultimately, for the private sector, the choice to not provide environmental public goods is a dominant strategy. In the case that the private sector is unwilling to provide environmental public goods, the government has to choose to bear all the costs

[4] Chen Jing: Considerations of Perfecting China's Environmental Protection Law, Journal of Xinjiang University (Philosophy, Humanities & Social Sciences), 2005, No. 4.

required for providing such goods. This will cause the government to be overburdened by the investment in environmental public goods, and lead to over-high financial expenditure on the project. Another parallel is that it restricts the government's investment in the development of other industries, thereby causing the allocation of social resources to be incapable of reaching Pareto optimality.

As regards the above "Boxed Pigs Game" between the government and the private sector in the provision of environmental public goods, it does not mean that there is no solution. According to Table 5.2, when the government and the private sector jointly provide environmental public goods, both of them can obtain certain revenue. How do we achieve this ideal state? The government may show its choice clearly. In other words, in case the private sector is willing to provide environmental public goods, the government may choose cooperation, rather than sit idle and enjoy the fruits. If the private sector is unwilling to undertake part of the supply, the government may choose not to provide either. Then, their revenue is both 0. Obviously, after understanding the government's intention, the private sector will choose to provide part of environmental public goods for its own sake. In the process, the credibility of government commitments plays a decisive role. If the government sector chooses not to participate in the supply of environmental public goods after making a commitment to cooperate with the private sector on supply, the private sector will no longer choose to provide such goods. To increase the credibility of government commitments, one method is to enable the government to publicize its commitment, and allow the entire society to supervise its behavior as supervisors; the other method is to accurately specify the proportions to be respectively undertaken by the government sector and the private sector in the supply process of environmental public goods in legal provisions, thereby imposing a binding constraint on the behaviors of both sectors, and making them both highly credible.

Chapter 5
Feasibility Research on Residents' Participation in the Supply of Environmental Public Goods

Theoretically, residents' participation in environmental governance has the effect of improving their environmental benefits, and the improvement of environmental quality can promote the health of future generations and elevate the human capital thereof [129]. Well, do Chinese residents have such environmental awareness and responsibility? Are residents subjectively willing to pay pollution control fee?

5.1 Analysis on Residents' Subjective Willingness to Participate in Supply of Environmental Public Goods

For building a mechanism for government-led voluntary supply of environmental public goods by residents, the crucial factor is residents' subjective willingness for environmental protection: Are they willing to pay? How should we determine a payment standard? Therefore, we use related data from the Chinese General Social Survey (CGSS) for empirical analysis so as to determine the degree of residents' concern for environmental pollution, and their willingness to pay for environmental improvement (e.g. pay taxes).

5.1.1 Residents' Growing Concern for Environmental Pollution

At present, environmental pollution has become a problem that affects people's wellbeing and sustainable development, and the masses have made a strong appeal for improving the ecological environment. According to the data of CGSS 2010, nearly 70% of residents showed concern for environmental problems. Among them,

J. Yang, *Effective Supply Mechanism and Path of Environmental Public Goods in China*, Public Economy and Urban Governance in China,
https://doi.org/10.1007/978-981-99-2939-9_5

Table 5.1 Degree distribution of residents' concern for environmental problems

CGSS (a)					CGSS (b)	
Question	Generally, how concerned are you about environmental problems?		According to your own judgment, generally, do you think the environmental problems faced by China are serious?		Do you think these environmental problems are serious in your region?	
Options and statistical results	Totally unconcerned	3.10	Very serious	21.16	Quite serious	17.34
	Relatively unconcerned	10.65	Relatively serious	48.42	Relatively serious	24.48
	Falling in between	19.23	Falling in between	10.43	Less serious	17.44
	Relatively concerned	48.31	Less serious	11.27	Not serious	17.63
	Very concerned	17.16	Not serious at all	0.68	Ordinary	9.99
	Unable to choose	1.42	Unable to choose	7.63	Unconcerned/ unexplainable	0.81
	Do not know/ refuse to answer	0.14	Do not know/ refuse to answer	0.41	N/A	12.30
Sample	Sample size	3672	Sample size	3672	Sample size	4787

Data source (a) Some data was calculated according to the data of CGSS 2010; (b) some data was calculated according to the data of CGSS 2013
Unit %

48.31% of residents chose "relatively concerned"; 48.42% of residents considered that China's environmental problems were "relatively serious" while 21.16% of residents chose "very serious". Both the central and local governments increased efforts to govern the environment. According to data: in 2013, China increased investment by 124.89% in curbing waste gas over the previous year; East China increased investment by 82.19% therein over 2012; West China increased investment by 176.78% therein over 2012.[1] However, according to the survey results of 2013, 41.82% of respondents considered that environmental problems were "very serious" or "relatively serious"; in the survey of 2014, 69.58% of respondents chose the two options (Table 5.1). This raised a new requirement for the government's capability of environmental governance. How to take effective measures for environmental governance has become a focus of national attention.

[1] Calculated according to the data from the website of the National Bureau of Statistics.

To find out which types of pollution have made a deep impression on residents, this question was surveyed in the CGSS questionnaire of 2013. "B21 Do you know the following types of environmental problems?" 12 options are listed below the question—"Air pollution, water pollution, noise pollution, food pollution, industrial pollution, household waste pollution, insufficient green space, destruction of forest vegetation, degradation of farmland quality, shortage of fresh water, desertification and reduction of animals and plants". The percentages of respondents choosing "Yes" included: 90.13% for air pollution; 89.76% for water pollution, 87.13% for household waste pollution, and 81.4% for noise pollution. See Table 5.2 for the specific data.

The statistical data of the survey suggested that the top five environmental problems that residents deemed relatively serious in their region were air pollution, desertification, household waste pollution, water pollution, and noise pollution. See Table 5.2 and Fig. 5.1 for details.

The same questions were presented in the 3672 sample survey questionnaires of CGSS 2010. "Which one do you think is the most serious environmental problem in China?" "Which one do you think has the largest impact on you and your family?" The top three were respectively air pollution, water pollution and household waste pollution, which produced the largest impact on families in the view of residents. See Table 5.3 for specific data.

In Fig. 5.2, pollution sources are sorted by the degree of their influence on families from the perspective of residents. It is evident that air pollution, water pollution and household waste pollution closely connected with the ordinary people's livelihood and even lives have become residents' biggest concern. Air pollution seriously affects our physical health. Scholars' empirical researches show that SO_2 pollution is not only highly related to infant mortality [130, 131], but also highly associated with adult mortality. Schwartz and Marcus [132] analyzed the time series data of London from 1958 to 1972; Mendelsohn and Orcutt [133] researched the cross-section data of the United States in 1970, and drew the above conclusions as well. Through an empirical research, Chen et al. [5] found that the TSP pollution in North China was 55% higher than the national average, thus residents' life expectancy was shortened by 4.5 years.

Faced with such severe environmental problems, how did residents respond to them? Table 5.4 makes statistics on the environmental protection actions taken by residents. Only 19.23% of residents took action, but 43.36% of residents did not take any action. This implied that nearly half of residents just passively accepted pollution. There is another noteworthy datum. 23.99% of respondents chose "Tried to take action, but didn't know what to do". In other words, a significant number of residents expected environmental protection actions, but needed professional guidance or training.

Table 5.2 Frequency distribution of residents' subjective feelings for environmental pollution

Option	Do you know the following types of environmental problems			Do you think these environmental problems are serious in your region?							
Pollution item	Yes	No	Refuse to answer	Very serious	Relatively serious	Less serious	Not serious	Ordinary	Unconcerned/ unexplainable	N/A	Refuse to answer
Air pollution	90.13	9.84	0.03	17.34	24.48	17.44	17.63	9.99	0.81	12.30	0.04
Water pollution	89.76	10.20	0.03	12.76	24.86	19.74	17.19	10.36	1.88	13.14	0.06
Noise pollution	81.40	18.56	0.03	12.03	21.27	20.22	17.40	11.82	1.50	14.71	0.04
Industrial pollution	76.74	23.21	0.05	8.79	18.82	14.75	16.15	11.01	3.72	24.69	0.06
Household waste pollution	87.13	12.83	0.04	11.11	28.18	18.30	16.42	13.27	1.50	11.13	0.08
Insufficient green space	66.19	33.76	0.05	8.67	19.26	13.60	14.17	16.00	4.79	21.43	0.08
Destruction of forest vegetation	66.84	33.14	0.03	6.71	13.75	10.86	14.04	12.24	8.54	33.82	0.04
Degradation of farmland quality	64.99	34.97	0.04	8.31	19.55	10.07	11.61	10.76	11.30	28.31	0.08

(continued)

Table 5.2 (continued)

Option	Do you know the following types of environmental problems			Do you think these environmental problems are serious in your region?							
Pollution item	Yes	No	Refuse to answer	Very serious	Relatively serious	Less serious	Not serious	Ordinary	Unconcerned/ unexplainable	N/A	Refuse to answer
Shortage of fresh water	69.58	30.35	0.06	8.52	16.02	11.57	16.50	11.34	6.45	29.50	0.08
Desertification	74.06	24.87	0.07	17.46	22.12	11.57	13.06	10.82	6.43	18.40	0.13
Reduction of animals and plants	50.88	49.06	0.06	3.95	8.11	7.10	9.09	6.91	10.93	53.77	0.15
Sample size	11,438			4787							

Data source Calculated according to the data of CGSS 2013
Unit %

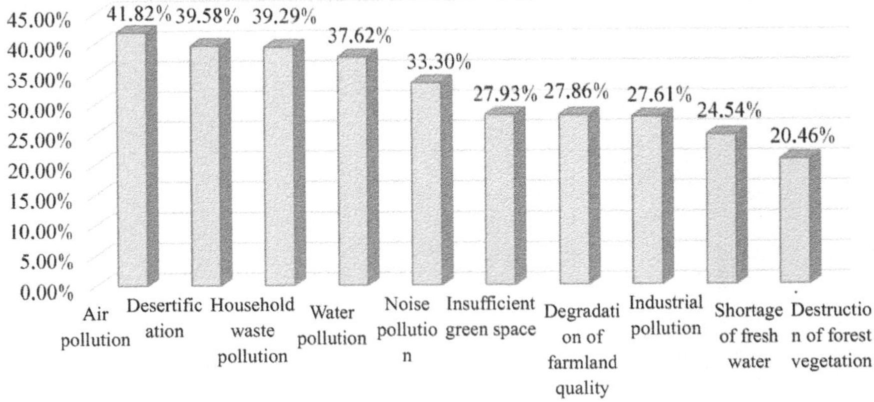

Fig. 5.1 Residents' subjective feelings for environmental pollution. *Data source* Drawn upon calculation according to the statistical results of CGSS 2013; the sum of the percentages of respondents choosing the two options "very serious" and "relatively serious" was used for sequenced comparison

Table 5.3 Residents' subjective feelings for environmental pollution

Option	Which one do you think is the most serious environmental problem in China?		Which one do you think has the largest impact on you and your family?	
	Number of respondents	Percentage (%)	Number of respondents	Percentage (%)
Air pollution	1119	30.47	845	23.01
Water pollution	652	17.76	615	16.75
Household waste pollution	581	15.82	694	18.9
Fertilizer and pesticide pollution	328	8.93	497	13.53
Climate change	187	5.09	126	3.43
Shortage of water resources	170	4.63	176	4.79
Depletion of natural resources	131	3.57	30	0.82
GMF	32	0.87	59	1.61
Nuclear waste	18	0.49	3	0.08
Refuse to answer	9	0.25	5	0.14
N/A	2	0.05	4	0.11
None of the above	35	0.95	253	6.89
Unable to choose	408	11.11	365	9.94
Sample size	3672			

Data source Obtained from the survey results of Question 17a/b in CGSS 2010

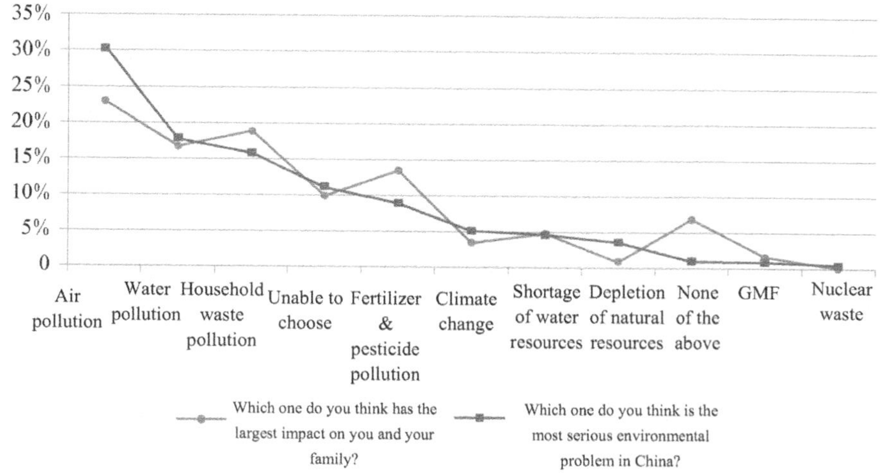

Fig. 5.2 Residents subjective feelings for local pollution. *Data source* Obtained from the survey results of question 17a/b in CGSS 2010

Table 5.4 Environmental protection actions taken by residents

To solve the environmental problems faced by you and your family, did you take any action?		
Option	Number of respondents	Percentage (%)
Did not take action	1592	43.36
Tried to take action, but didn't know what to do	881	23.99
Took action	706	19.23
No environmental problems	474	12.91
Others	19	0.52

Data source Obtained from the survey results of CGSS 2010

The survey probes into residents' perceptions of environmental pollution and solutions. Table 5.5 makes detailed statistics. Only 28.59% of them had some understanding of the causes of environmental pollution. For other residents, some had little understanding, and some were confused. Besides, 12.2% of them chose "Have no understanding". As regards solutions to environmental pollution, only 14.68% of residents chose "Have some understanding", but most of them were confused or knew nothing.

The above surveyed residents' perceptions of environmental pollution, understanding of the causes thereof and solutions thereto suggested that residents had some understanding of severe pollution problems, but only few of them took practical action, and needed professional knowledge training and guidance for solutions.

Table 5.5 Residents' mastery degree of causes of environmental pollution and solutions thereto

Option	How much do you understand about the causes of the above environmental problems?		How many solutions do you have regarding the above problems?	
	Number of respondents	Percentage (%)	Number of respondents	Percentage (%)
Have little understanding	1335	36.36	1617	44.04
Have some understanding	1050	28.59	539	14.68
Confused	586	15.96	618	16.83
Have no understanding	448	12.2	676	18.41
Unable to choose	155	4.22	180	4.9
Have an in-depth understanding	87	2.37	31	0.84
Others (do not know, N/A)	11	0.3	11	0.3

Data source Obtained from the survey results of CGSS 2010

5.1.2 *Residents with Certain Willingness to Pay for Environmental Improvement*

Residents recognized that current environmental pollution was serious. Did they have a strong subjective willingness to pay for environmental improvement? In the CGSS questionnaire of 2010, there are several items related thereto. First, "To what extent are you willing to pay higher prices for environmental protection?" The question is arranged with six options: "Strongly willing; relatively willing; neither willing nor unwilling; less willing; strongly unwilling; and unable to choose". Second, "To what extent are you willing to pay higher taxes for environmental protection?" The listed options are identical to that of the previous question. In Table 5.6, we displayed the survey results of the above two questions. It was gratifying that only 7.05% and 8.71% of respondents answered "strongly unwilling". Although only 33.82% (residents relatively willing to pay higher prices for environmental protection) and 28.35% (residents relatively willing to pay higher taxes for environmental protection) answered "relatively willing", we saw that nearly 10% of respondents chose "strongly willing", and neutrals accounted for 18.33% and 18.85% respectively. If these neutrals could be brought into the team willing to pay, the percentages of people answering "willing" to the above questions would reach 60.65 and 52.78%, both more than half. Additionally, the percentage of respondents subjectively willing to pay higher prices for environmental protection exceeded that of respondents subjectively willing to pay taxes. We could see that residents had a strong subjective willingness to

Table 5.6 Statistics on residents' willingness to pay for environmental improvement

Option	To what extent are you willing to pay higher prices for environmental protection? (%)	To what extent are you willing to pay higher taxes for environmental protection? (%)
Strongly willing	8.50	5.58
Relatively willing	33.82	28.35
Neither willing nor unwilling	18.33	18.85
Less willing	23.01	27.89
Strongly unwilling	7.05	8.71
Unable to choose	8.85	10.19
Missing value	0.43	0.38
Sample size	3672	

Data source Calculated according to the data of CGSS 2010

participate in environmental protection, which laid a solid foundation for motivating the entire people (mostly residents) to participate in environmental protection.

The above research suggests that with the acceleration of economic growth and the improvement of residents' living standards, residents have continued to elevate their willingness to pay for environmental improvement. This has corroborated the following views proposed by a scholar: From a social perspective, when certain material conditions are satisfied, residents will continue to improve their willingness to pay for environmental improvement; income level, degree of participation in environmental protection, and educational attainment are directly correlated to residents' willingness to pay; besides, younger residents will have a higher willingness to pay for the environment [134].

5.2 Assessment Methods of Residents' Willingness to Pay for Environmental Protection

In terms of participation in environmental governance, residents pay for it on a voluntary basis, but how to set a benchmark for pollution control fee under the guidance of the government is an important yet difficult problem. Only after the problem is solved can the mechanism of joint governance by the government and residents function effectively and have a scientific evaluation criterion.

As regards the measurement of residents' willingness to pay for environmental protection, namely the valuation methods of public goods, the following three research ideas for valuating public goods proposed in the academic circles can be used for reference: the revealed preference method, the stated preference method and the happiness measurement method.

5.2.1 Revealed Preference Method

The basic principle of the revealed preference method is inferring consumers' preferences according to their purchase behavior, mainly because consumers' purchase behavior under certain price conditions betray or reveal their preferences. Therefore, we compare the asset values in different environments to infer residents' preferences for certain public goods.

One of the common methods is the hedonic pricing method, also known as hedonic pricing model method and utility valuation method. According to this method, the price of a commodity is composed of different characteristics or attributes in multiple aspects. For example, in addition to housing type, area and orientation, the price of real estate is also influenced by the surrounding ecological environment. We can decompose the factors influencing the price of real estate to obtain the price implicit in influencing factors. If the number of characteristics (or attributes) of real estate remains constant under control, the attribute factor of changes in real estate prices can be detached so as to purely reflect the changes in prices.

In this regard, scholars carried out some significant researches. For instance, Briscoe's [135] research on the price of real estate in Brazil suggested that the clean-up of dumps significantly increased the local housing price and rent, and the aggravation of pollution significantly lowered the housing price. Wang et al. [136] employed similar methods to research residents' willingness to pay for air pollution control in Qingdao, and their research results showed that each household was willing to pay 600 Yuan/year.

Another method is the transport cost method, which researches the people's transport cost for consuming public healthcare services to determine the degree of their preference for public healthcare services [137]. Feng and Lu [138] probed into the effect of school choice behavior on the real estate market of Shanghai. Other methods including the protective expenditure method and the travel cost method are also the application of the revealed preference method. Zeng et al. [139] measured residents' willingness to pay for the decline of different concentrations of pollutants by researching residents' perception of $PM_{2.5}$ health risk in Beijing.

5.2.2 Stated Preference Method

The stated preference method is a direct preference revelation mechanism. Questionnaire survey is a common method thereof, which mainly reflects the people's real willingness to pay for public goods in the form of questionnaires, i.e. getting to know how much money respondents are willing to make up for the hypothetical gains or losses of non-market goods and services by means of questionnaires. After the 1980s, the stated preference method received extensive attention. Scholars learned about residents' willingness to pay for stable water supply, and their willingness to pay for health risk reduction by this method.

Li et al. [140] employed the stated preference method to survey the residents of Chengdu, and concluded by calculation that residents were willing to pay RMB 234.04 averagely every year. Wei et al. [141] employed the contingent valuation method (one of the stated preference methods), i.e. directly surveying and consulting about the people's willingness to pay (WTP) for ecological services, used WTP and net willingness to pay (NWTP) to express the economic value of environmental goods—evaluated the value of ecological protection and construction of Yanqing District—an ecological conservation area in Beijing with regard to the improvement of Beijing's air quality, and concluded that each household was willing to pay RMB 283.91-404.34 every year.

The above two methods both have some deficiencies. Based on subjective judgments, the revealed preference method has many important variables which can be hardly measured, causing researchers to be incapable of obtaining non-use value, or assessing utility losses. The stated preference method reflects the respondents' subjective value, thus unreliable results may arise from the insensitivity of individual WTP to the supply quantity and scope of public goods. Moreover, unfamiliar with public goods, the respondents are less likely to consider the budget constraints and substitution relations when answering questions.

5.2.3 Happiness Measurement Method

The idea of happiness measurement came from Easterlin [142], who first associated residents' income with happiness for research. Later, more and more people in the academic circles began to research the relations between the two. Some scholars found that the increase of income could significantly boost happiness. As the call for environmental protection has grown louder, recent researches have begun to focus on the effect of pollution and income on residents' happiness. Lots of researches suggest that pollution significantly lowers residents' happiness while income increases their happiness. After knowing the marginal utility of income and pollution, we can measure the marginal rate of substitution of pollution for income so as to learn about how much money residents are willing to pay for every 1 unit of pollution decrease, i.e. willingness to pay. Using the LSA method, scholars empirically researched residents' willingness to pay for noise pollution, hurricanes, air pollution, floods, terrorist attacks, and so forth [143].

The happiness measurement method was proposed by German scholar Welsch [20]. This method assumes that subjective happiness (well-being) is a good proxy variable of individual utility level. Besides, the happiness decision function can be expressed as:

$$H = H(P, Y, X) \tag{5.1}$$

where H denotes subjective happiness, used for measuring individual utility level; P denotes the degree of environmental pollution; Y denotes income level.

In Eq. (5.1), we assume that P and Y respectively have negative and positive effects on happiness, i.e. $\partial H/\partial P < 0$ and $\partial H/\partial Y > 0$. The above first-order conditions have been supported by numerous empirical researches. Some scholars found that such air pollutants as SO_2, NO_2 and suspended particulates all significantly lowered residents' happiness [153]; additionally, seen from the cross-section, high income earners had higher happiness than low income earners. X denotes other factors influencing happiness, and is deemed as an exogenous variable. The perfect differential of the above function is set as $\mathrm{d}H = 0$, i.e. $\mathrm{d}H = (\partial H/\partial P)\,\mathrm{d}P + (\partial H/\partial Y)\,\mathrm{d}Y = 0$. Then, the marginal rate of substitution between environmental pollution and income level, i.e. representative individuals' marginal willingness to pay (MWTP) for environmental pollution, can be obtained.

$$\mathrm{MWTP} = \mathrm{d}Y/\mathrm{d}P = -(\partial H/\partial P)/(\partial H/\partial Y)\ldots \tag{5.2}$$

According to an intuitive interpretation of Eq. (5.2), individuals are willing to pay—$(\partial H/\partial P)/(\partial H/\partial Y)$ (amount of money) for reducing 1 unit of environmental pollution, in an attempt to maintain constant utility level. If we further assume that the happiness decision function is linear and serves as the logarithmic function of income, then it can be simplified into a linear equation easy to estimate:

$$H = \alpha P + \beta \ln Y + X'\gamma \tag{5.3}$$

Based on Eqs. (5.2) and (5.3), the average marginal willingness to pay can be obtained:

$$\mathrm{MWTP} = -Y(\alpha/\beta)\ldots \tag{5.4}$$

where Y denotes the sample mean of income level; parameters α and β can be obtained in the above linear equation by specific sample estimation.

According to the above content, $\alpha < 0$ and $\beta > 0$, thus MWTP > 0. If income level Y (rather than its logarithm) directly enters the happiness decision function, then marginal willingness to pay $\mathrm{MWTP} = \alpha/\beta$.

A basic assumption of the above happiness measurement method is that the subjective happiness obtained by survey is a good proxy variable of individual utility level. In the view of Levinson [22], subjective happiness is close to the "Experience Utility" proposed by Kahneman [145, 146]; individuals with a higher utility level tend to have higher happiness. However, it should be indicated that the classic choice theory is based on the theory of ordinal utility. Thus, even if happiness can be used to measure the utility level, different individuals' happiness may be incomparable. For this controversy, Ng [147] believed that ordinal utility was a forced assumption, and a lot of evidence had shown that utility was a cardinal number and interpersonally comparable. Furthermore, the empirical research on happiness tends to assume that happiness is an ordinal number but interpersonally comparable.

5.3 Residents' WTP for Environmental Protection Measured by Happiness Measurement

Theoretically, residents' marginal willingness to pay (MWTP) for environmental public goods—the amount of money that consumers are willing to pay for reducing 1 unit of air pollution—is an important parameter for developing the optimal environmental policy, with an aim to measure the economic value of clean air [157].

5.3.1 Measurement Advantage of the Happiness Measurement Method

Air pollution, water pollution, waste pollution, etc. are typical public goods with negative externality, resulting in the difficulty in revealing the "price" that consumers are willing to pay by means of market transactions. Thus, how to infer this parameter has always been a focus of attention in environmental economics.

At present, the protective expenditure method, the travel cost method, the hedonic pricing method, the contingent valuation method, etc. have been successfully applied to research, which deduce the people's valuation on environmental quality based on their market conduct and payout. However, the protective expenditure method only considers the direct cost incurred from environmental pollution, neglects its indirect cost (e.g. damages to landscape), and thus undervalues environmental quality; the travel cost method is only applicable to tourist attractions, and has a high implementation cost; the hedonic pricing method relies on the existence of Walrasian general equilibrium, but ignores information asymmetry, migration cost and other market imperfection factors. The latter mainly includes the contingent valuation method, which requires respondents to directly give economic assessments by positing an environmental change. This method is widely applicable, but likely to result in biased or inaccurate estimations by placing the respondents in an imaginary alien environment.

Compared to the above methods, the happiness measurement method deduces the "price" that individuals are willing to pay by directly observing the utility level (subjective happiness), overcomes the theoretical obstacle that public goods have no market prices because of no market transactions to some extent, and thus appears more direct than the "revealed preference method"; meanwhile, this method does not require respondents to directly undertake environmental quality assessment, and only needs a little information concerning respondents' subjective happiness and local environmental quality, so it is more indirect than the stated preference method.

In short, the happiness measurement method is an important supplement to traditional methods of environmental quality assessment.

5.3.2 Residents' MWTP for Environmental Public Goods Measured in Existing Researches

In existing researches, scholars [149–152] estimated the MWTP for air pollution in developed countries including the UK and the US, i.e. the amount of money that consumers are willing to pay for reducing 1 unit of air pollution, which is an important parameter for developing environmental policies. Scholars not only used this method to estimate the MWTP for air pollution in such developed countries, but also extended the above method to the evaluation on noise [2], floods and terrorism [153]. So far, China has few case studies in this respect. Yang and Zhang [154] applied the happiness measurement method to estimate Chinese residents' MWTP for air pollution, but they only used the cross-section data of a single year, neglected weather, which is an important variable influencing both happiness and air pollution, and failed to consider the endogenous problem of air pollution. Thus, it is difficult to determine the bias or inaccuracy degree of their estimations.

These researches calculate the amount of fee that residents are willing to pay for pollution based on the happiness measurement method. Table 5.7 displays residents' MWTP for participation in environmental governance of different countries in existing researches, providing a calculation basis for China's policy introduction and the payment for pollution control fees in regions with imbalanced economic development.

Carlsson et al. [155] researched Swedish residents' WTP for air control, and estimated that their average WTP was SEK 2000 every year. Gao et al. [156] researched Beijing and Lanzhou residents' WTP respectively, and estimated that Beijing residents' average WTP for lowering air pollutant concentration by 50% within five years was RMB 652.33 every year; Lanzhou residents' average WTP for level-2 air quality was RMB 140.97 (household/year). Besides, through research, He and Huang [157] estimated that to reduce 1 pollution day (inferior to the urban level-2 standard), residents were willing to pay RMB 353.41 every year; each household was willing to pay RMB 2473.88 every year for reducing 1 unit of soot pollution and RMB 2148.55 for alleviating 1 kg SO_2 pollution. Zeng et al. [139] estimated that in case of lowering the concentration of fine particulates ($PM_{2.5}$) by 60%, residents' average WTP was RMB 39.82 every month.

Table 5.7 Residents' MWTP for air purification and other disaster prevention and control

Pollutant	MWTP (Yuan/year)	Percentage of income[2] (%)	Estimation method	Sample and time span
PM_{10}	834		Ordinary least squares (OLS)	Beijing, Tianjin, Shanghai and Chongqing, CGSS pooled data from 2003 to 2013
	1784		Instrumental variables (IV)	
Pollutant	MWTP/(Yuan/year)	Percentage of income (%)	Estimation method	Sample and time span
NO_2	1005	2.71	Ordinal logit	China, 2010
	1250	3.37	OLS	
	1144	3.09	Ordinal probit	
SO_2	2148		OLS	China, CGSS 2010
Soot	2473			
PM_{10}	6031		OLS	The US, from 1986 to 1996
	6024		Ordinal probit	
	4927		IV	
SO_2	USD 150	0.6	OLS	13 European countries, from 1979 to 1994
	USD 312	1.1	OLS (plus IV)	
	USD 154	0.6	Pooled ordinary least squares (POLS)	
	USD 344	1.3	POLS (plus IV)	
Floods	USD 6399	23.3	Ordinal probit	16 European countries, from 1973 to 1998
	USD 6505	23.7	OLS	
Terrorism		4.2 4	OLS Ordinal probit	France, from 1973 to 1998
		8.2 7.5	OLS Ordinal probit	
		24.6 26.6	OLS Ordinal probit	British Isles, from 1975 to 1998
		37.3 38.4	OLS Ordinal probit	

(continued)

[2] Here income level is a core explanatory variable. According to existing literature, most scholars used household income as the proxy variable of income level. Thus, the willingness to pay (WTP) in the above table refers to household WTP, rather than individual WTP.

Table 5.7 (continued)

Pollutant	MWTP (Yuan/year)	Percentage of income (%)	Estimation method	Sample and time span
SO_2		0.9–1.5	LSA	Germany, from 1985 to 2003
		0.03–0.2	Hedonic pricing model	
NO_2	USD 760		OLS	10 European countries, from 1990 to 1997
Pb	USD 1390		OLS	
NO_2	USD 70/KT		OLS	54 Countries

Chapter 6
Extended Analysis on Factors Influencing Residents' Participation in the Supply of Environmental Public Goods

According to the CGSS report, nearly half of the residents surveyed were willing to pay for environmental public goods, and obviously, a significant number of them were "free-riders". To motivate more residents to participate therein as the supply side of quality environment, we must have a clear understanding of the factors influencing their WTP in addition to the thought of "free-riding".

In existing researches, Levinson [22] used the data of the US General Social Survey and the air quality system data of the US Environmental Protection Agency to measure and calculate residents' WTP for lowering the concentration of PM_{10} in air, and found that higher income groups were more willing to pay for reducing pollution; groups living in places with better air quality, healthier groups and higher-educated groups were more willing to pay more for reducing pollution.

As regards the researches of China, Li et al. [158] used the contingent valuation method to survey Beijing residents' WTP for improving ambient air quality, and the survey results suggested that WTP increased with the rise of income level and educational attainment, and decreased with the rise of age and family size. Improving income level and enhancing environmental awareness is an effective approach of improving public WTP. Upon a regression analysis on the pollution in 86 cities of China in 2009 and a series of influencing factors thereof, Zheng [159] found that cities with higher human capital where residents were more concerned about the environment could reach the turning point of the Environmental Kuznets Curve (EKC) earlier. More developed coastal areas had stricter environmental regulation policies, and urban residents had a stronger desire to pursue higher environmental quality, which spurred residents to have a higher WTP for pollution control. Yang and Zhang [124] employed the LSA method to empirically research Chinese residents' WTP for air pollution, and found that air pollution had a larger effect on low income

J. Yang, *Effective Supply Mechanism and Path of Environmental Public Goods in China*,
Public Economy and Urban Governance in China,
https://doi.org/10.1007/978-981-99-2939-9_6

groups, men and rural residents, and Chinese residents' WTP for air pollution was significantly lower than that of Western developed countries. Upon research, Mu and Fan [160] found that the cognition degree of knowledge about air pollution had significantly positive correlations with residents' WTP. Residents having a better understanding of air pollution knowledge could more easily learn about the harmfulness of air pollution and the changes in individual benefits after environmental improvement, and were more willing to pay fees for air pollution control. Besides, for residents "having an in-depth understanding" of air pollution knowledge, their probability of being willing to pay for air pollution control was 7.72 times higher than that of residents "having no understanding".

6.1 Effect of Material Life on Residents' Participation in Environmental Governance

In general, the EKC is often used to research the relations between environmental quality and economic development. With the improvement of the economic development level, the people may raise their demand for environmental quality improvement. In the view of Peng and Bao [161], with the improvement of the economic level, the people are more willing to choose environment-friendly or green products, which will encourage enterprises to transform their production mode to clean production and reduce pollutant discharge. In the view of Zhang and Zhang [162], technological progress, structural improvement, as well as other public or private control measures concerning consumption, production, systems, and pollutant discharge reduction can be motivating factors of environmental quality improvement.

Rostow's [163] economic development theory holds that the six stages of economic development include the traditional society, the preconditions for take-off, the take-off, the drive to maturity, the age of high mass-consumption, and beyond consumption. The people's life quality improvement needs to undergo four stages, respectively poverty, take-off, high material consumption, and search for quality (quality of life). In the view of Rostow, "take-off" and "search for quality" are two critical stages. In the book *Politics and the Stages of Growth*, he mentioned the stage of "search for quality", the leading sectors of which were the service sector and the environmental renovation business sector. Usually, in latter stages as described by Rostow, outdoor consumption, public goods and service consumption appear increasingly important; the people are more concerned about ecological environment; and public goods occupy a larger weight in the utility function.

6.1.1 *Effect of Income Level on Residents' Willingness to Participate in Environmental Protection*

What stage of development are we in? How is the masses' life quality in China? If the masses' life quality in China is in the latter stages of development as described by Rostow, they will naturally raise their concern for environmental protection, and more people will voluntarily join the environmental protection team.

This book applies gross domestic product (GDP), per capita disposable income and Engel's coefficient to measure the above development stage issues. Engel's coefficient is an important indicator for measuring the people's living standards of a country or region. In a poorer country (or region), the expenditure on food will account for a greater percentage of nationals' average income (or average expenditure). As the country (or region) grows richer, the percentage will show a downward trend.

According to Engel's coefficient, the United Nations has a dividing criterion for the living standards of countries around the world, i.e. for a country, the average household Engel's coefficient above 60% denotes poverty; 50–60% denotes enough food and clothing; 40–50% denotes moderate prosperity; 30–40% denotes a well-to-do living standard; 20–30% denotes a wealthy living standard; below 20% means a fairly wealthy living standard.

In 2016, the per capita disposable income of Chinese residents was RMB 23,821, increasing by RMB 7311 over 2012, with an average annual real growth of 7.4%. In the first half of 2017, residents' per capita disposable income had a real growth of 7.3% on year-on-year basis, which was 0.4% higher than the growth rate of GDP, and 0.9% higher than that of per capita GDP. Residents have accelerated their pace of consumption upgrading.

Table 6.1, Figs. 6.1 and 6.2 compare the material life conditions and life quality of Chinese urban and rural residents from 2006 to 2019. The changes in material living standards are obtained upon a vertical comparison of per capita GDP; residents' life quality is analyzed by use of Engel's coefficient.

Figure 6.2 suggested that China's Engel's coefficient showed a downward trend. According to the data of recent years, China made significant progress in improving people's livelihood, and the development achievements thereof benefited the entire people, which were specifically manifested in the continuous improvement of residents' living standards. In 2016, Chinese residents' Engel's coefficient was 30.1%, dropping by 2.9% compared to that in 2012, which was close to the wealthy standard of 20–30% defined by the United Nations.

Table 6.1 Household income per capita of Chinese urban and rural residents and Engel's coefficient thereof[1]

Year	Per capita real income of urban residents (RMB per capita)	Per capita real income of rural residents (RMB per capita)	Urban Engel's coefficient (%)	Rural Engel's coefficient (%)
2006	11,760	3587	35.8	43
2007	13,154	3951	36.3	43.1
2008	14,902	4495	37.9	43.7
2009	17,296	5190	36.5	41
2010	18,499	5730	35.7	41.1
2011	20,692	6620	36.3	40.4
2012	23,942	7716	36.2	39.3
2013	25,796	9191	30.1	34.1
2014	28,278	10,283	30	33.5
2015	30,764	11,264	29.7	33
2016	32,957	12,121	29.3	32.2
2017	35,788	13,260	28.6	31.2
2018	38,444	14,316	27.7	30.1
2019	41,205	15,524	27.6	30.0

Data source The National Bureau of Statistics. The per capita income of urban and rural residents was calculated based on the price index 100 of 2006

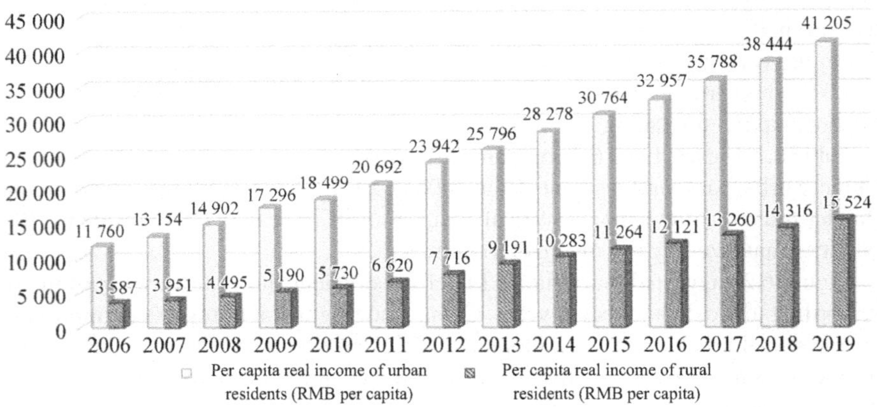

Fig. 6.1 Per capita income of Chinese urban and rural residents. *Data source* The National Bureau of Statistics. The per capita income of urban and rural residents was calculated based on the price index 100 of 2006

[1] Income and expenditure data of urban and rural residents before 2013 came from the urban and rural household sampling surveys carried out respectively. Since 2013, the National Bureau of Statistics has carried out the Integrated Household Survey of Income, Expenditure and Living Conditions. The data of 2013 and thereafter came from this survey. Compared to the urban and rural household surveys before 2013, this survey has a different survey scope, survey method, and indicator coverage.

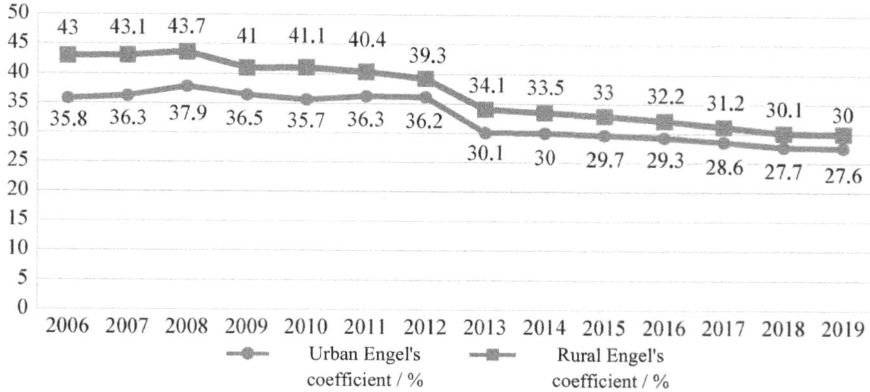

Fig. 6.2 Engel's coefficients of Chinese urban and rural residents. *Data source* The National Bureau of Statistics

From 2013 to 2016, China's GDP had an annual average growth of 7.2%, higher than the global average of 2.6 and 4% of developing economies; China contributed an average of 30% to the global economic growth, exceeding the sum of contribution rates of the US, the Eurozone and Japan; in 2017, China's per capita gross national income (GNI) rose to over USD 8000 (calculated according to the exchange rate of RMB to USD, 675.18 Yuan to 100 US dollars), close to the average level of upper middle-income countries. In 2019, China's per capita GNI exceeded USD 9000 (calculated according to the exchange rate of RMB to USD, 689.85 Yuan to 100 US dollars). It could be observed that the Chinese people's living standards were on a rapid rise (Fig. 6.3).

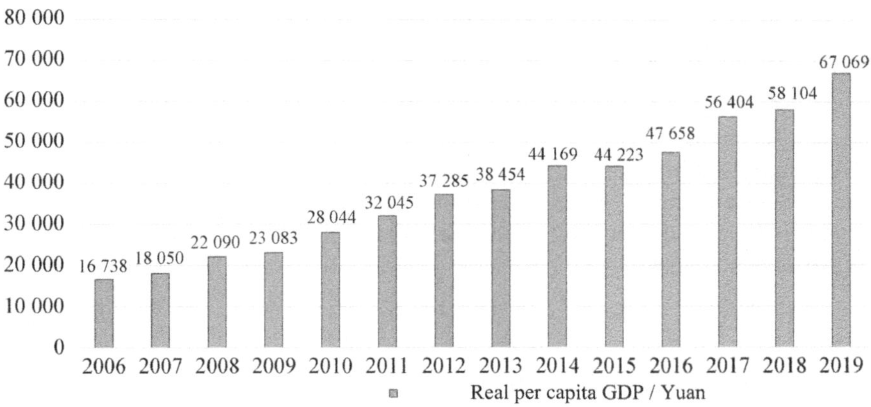

Fig. 6.3 Per capita GDP of China

6.1.2 Effect of Technological Progress on Residents' Willingness to Participate in Environmental Protection

Many scholars focus on researching the effect of technological progress on carbon emission reduction [164]. After conducting a field survey on 2000 Dutch families in 1999, Poortinga et al. [165] found that in order to save energy, high income families preferred to improve technology, rather than change behavior style.

In fact, we found that the continuous technological progress could not only directly influence the effect of discharge reduction by enterprises, but also facilitate the better compliance of residents with regulations on environmental protection. Take the waste problem that we are concerned about in life as an example. After making technological progress in waste sorting, we can not only complete harmless disposal, but also recycle part of the waste. These technologies could not only spur residents to focus more on and perform waste sorting in life, but also motivate them to take an active part in environmental protection actions.

Tables 6.2 and 6.3 and Fig. 6.4 sort out the household waste disposal rates of Chinese residents. Influenced by technological progress, the harmless disposal rates of household waste in all provinces (autonomous regions/municipalities) of China continuously increased; Shanghai and Shandong had realized a 100% harmless disposal rate for three consecutive years; in 2016, 27 provinces (autonomous regions/ municipalities) realized a harmless disposal rate of over 90%, and 29 provinces (autonomous regions/municipalities) realized a harmless disposal rate of over 80%.

6.2 Analysis on Residents' Environmental Awareness

In future China, the people will pursue a higher-quality life, e.g. better education, more stable jobs, higher income, more reliable social security, better medical and health care, more comfortable living conditions, a more beautiful environment, and a richer spiritual and cultural life. These needs are diversified, personalized, variable and multi-layered.

A beautiful environment needs the participation of all residents. If we could have a clear understanding of residents' environmental awareness and participation willingness, policies and measures would be more targeted in solving problems related to environmental governance.

6.2.1 Chinese Residents Strongly Dependent on the Government in Environmental Awareness

Environmental awareness is a combination of social thoughts, theories, emotions, willpower, consciousness and other ideologies reflected in the process of dealing

Table 6.2 Harmless disposal rates of household waste in 30 provinces (Autonomous Regions/ Municipalities) of China

Year	2008	2009	2010	2011	2012	2013	2014	2015	2016	2017	2018
National average	66	71	78	80	85	89	92	94	96	97	98
Beijing	98	98	96	98	99	99	100	79	100	100	100
Tianjin	94	94	100	100	100	96	96	93	94	94	95
Hebei	57	59	70	73	81	83	86	95	98	100	100
Shanxi	47	63	74	78	80	88	92	97	95	95	100
Inner Mongolia	55	72	83	83	91	94	95	98	99	99	100
Liaoning	60	60	71	80	87	88	92	94	93	99	100
Jilin	33	38	45	49	45	61	62	85	85	72	87
Heilongjiang	25	30	40	44	48	54	59	78	81	83	87
Shanghai	74	79	82	61	84	91	100	100	100	100	100
Jiangsu	91	91	94	94	95	97	98	100	100	100	100
Zhejiang	90	98	98	95	99	99	100	99	100	100	100
Anhui	54	61	65	86	91	99	100	100	100	100	100
Fujian	88	93	92	95	95	98	98	99	98	99	100
Jiangxi	80	84	85	88	89	93	93	94	95	98	100
Shandong	79	91	92	93	98	99	100	100	100	100	100
Henan	67	74	83	84	85	90	93	95	99	100	100
Hubei	53	55	61	61	72	84	90	92	95	100	100
Hunan	60	66	79	85	94	95	100	100	100	100	100
Guangdong	64	64	72	72	79	85	85	92	95	98	100
Guangxi	82	85	91	94	98	95	94	99	99	100	100
Hainan	65	65	68	91	100	100	100	100	100	100	100
Chongqing	88	95	99	100	99	99	99	99	100	99	100
Sichuan	81	84	86	88	88	95	94	96	99	99	99
Guizhou	76	82	91	89	92	92	93	94	95	95	96
Yunnan	80	81	88	74	83	88	92	90	93	93	98
Shaanxi	69	69	80	90	88	95	95	98	99	99	99
Gansu	32	32	38	42	42	42	63	64	73	98	100
Qinghai	74	64	67	89	89	78	85	87	95	95	96
Ningxia	55	42	93	66	71	93	93	90	98	99	99
Xinjiang	52	61	71	79	79	78	82	81	83	89	91

Unit %

Table 6.3 Comparison of harmless disposal rates of household waste

Year	2008	2009	2010	2011	2012	2013	2014	2015	2016	2017	2018
National average (%)	66	71	78	80	85	89	92	94	96	97	98
Number of provinces (cities) above the national average	16	16	16	19	19	19	21	18	17	22	23
Number of provinces (cities) with a rate above 90%	3	8	10	11	13	19	23	23	27	27	28

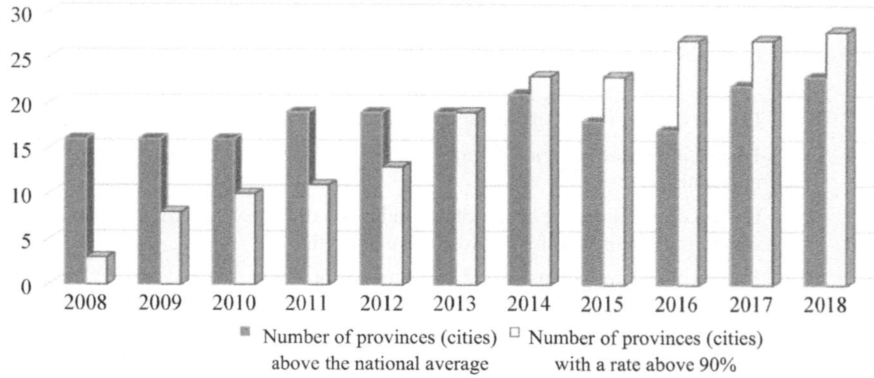

Fig. 6.4 Harmless disposal rates of household waste in 31 provinces (Autonomous Regions/ Municipalities) of China. *Data source* Former Ministry of Environmental Protection

with the relations between mankind and natural environment, and mainly manifested as environmental knowledge, attitudes and behaviors [166]. The formulation of our environmental protection and control measures should be based on the existing environmental awareness of the general public. Gao [167] used the CGSS data for regression analysis, and drew the following conclusion: As an environmental sense, the degree of residents' concern for environmental problems plays a positive role in promoting WTP for environmental protection; residents showing higher concern for the environment will have a stronger WTP for environmental protection. How is the environmental awareness of Chinese residents? The CGSS reports of 2010 and 2013 both mention residents' environmental awareness. To sum up, the environmental awareness of Chinese residents have the following significant characteristics.

1. Weak environmental awareness, individual environmental protection behavior only accounting for a small percentage

The CGSS2010 includes some questions relating to residents' environmental awareness. The survey has total of 3672 valid questionnaires. We choose the following questions and answers given by the respondents (see Tables 6.4 and 6.5).

The above answers suggest that residents definitely answering "Always" and "Often" have strong environmental awareness, but only account for less than 48.75% (with respect to the answers to Q4: "Do you often save or reuse water for environmental protection?"); and only account for less than 40% with respect to other questions. This fully shows that residents are weak in environmental awareness. Meanwhile, we could see another set of data: no waste sorting and recycling system in the domicile nearby (25.88%); no provision of fruits or vegetables free of chemical fertilizers or pesticides (24.54%).

Questionnaire B of CGSS2013 includes: "22. Can you tell us whether you have engaged in any of the following activities or behaviors over the last year?" Among the survey questions, the following questions are significantly related to residents'

Table 6.4 Questions relating to residents' environmental awareness of 2010

Question No.	Question description
1	Do you often cut back on driving for environmental protection?
2	Do you often sort glass, aluminum cans, plastic or newspaper, etc. for ease of recycling?
3	Do you often reduce the household consumption of oil, gas, electricity and other energy sources or fuels for environmental protection?
4	Do you often save or reuse water for environmental protection?
5	Do you often choose not to buy any products for environmental protection?
6	Do you often intentionally buy fruits and vegetables free of chemical fertilizers or pesticides?

environmental awareness: Q1: Waste sorting; Q2: Bring your own shopping basket or bag when shopping for daily necessities; Q3: Reuse plastic bags; Q4: Maintain trees and green space at your own expense. The survey of 2013 has 11,438 valid questionnaires.

The survey data of Table 6.6 suggests that respondents answering "often" to the question concerning reuse of shopping bags account for the highest percentage, 49.25%, followed by those answering "often" to the question concerning BYO shopping bags, accounting for 39.46%. As regards the above two questions, the respondents choosing the option "occasionally" account for above 30%; however, the respondents answering "often" to waste sorting and forest planting only account for a small percentage. Seemingly, residents' individual environmental protection behavior is not so active, but it does not mean that they are subjectively unwilling to protect the environment. The reason may be that conditions are unavailable or insufficient, e.g. being unable to find sorting bins.

2. Strong herd mentality, heavily dependent on the government for environmental governance

In the survey questionnaire, there is a question: "'My efforts to protect the environment are meaningless unless everybody does it'. Do you agree with this view?" The respondents' answers are described in Fig. 6.5. 63% of them chose "Agree", which meant that most residents had a herd mentality in participation in environmental protection. According to the survey data, 21% of residents chose "Disagree"; 10% were neutrals; 6% chose "Other". These numbers gave us hope. In other words, a significant number of residents started from themselves to take an active part in environmental protection.

Figure 6.6 shows the responsible parties for environmental problems in the eyes of residents. The survey questionnaire lists options including enterprises, the government, civil groups and individual citizens. According to the survey results, nearly half of the respondents (49.54%) held the government responsible; 28.51% of them

Table 6.5 Statistical analysis on environmental awareness of Chinese residents

Question	Always	Often	Sometimes	Never	Refuse to answer	Do not know	N/A	No provision/No recycling system in the domicile nearby	I have no car or am incapable of driving a car
Q1	1.96	3.43	8.91	4.77	0.44	0.08	0.44	–	78.98
Q2	11.93	19.72	23.75	17.35	0.33	0.03	0.03	25.88	–
Q3	9.86	22.41	39.79	25.77	0.71	0.22	0.25	–	–
Q4	17.21	31.54	33.82	15.94	0.35	0.11	0.03	–	–
Q5	7.30	15.69	41.07	33.61	0.35	0.74	0.25	–	–
Q6	5.70	14.96	28.13	23.86	0.33	0.25	0.25	24.54	–

Data source Calculated according to the data of CGSS 2010

Table 6.6 Statistics on environmental protection behavior of Chinese residents of 2013

Response options	Waste sorting	Bring your own shopping basket or bag when shopping for daily necessities	Reuse plastic bags	Maintain trees and green space at your own expense
Never	54.12	24.17	18.72	84.79
Occasionally	32.41	35.14	31.78	11.23
Often	12.27	39.46	49.25	3.81
Refuse to answer	0.05	0.07	0.11	0.05
Do not know	0.14	0.12	0.11	0.11
N/A	0.01	0.04	0.03	0.01

Data source Calculated according to the data of CGSS 2013
Unit %

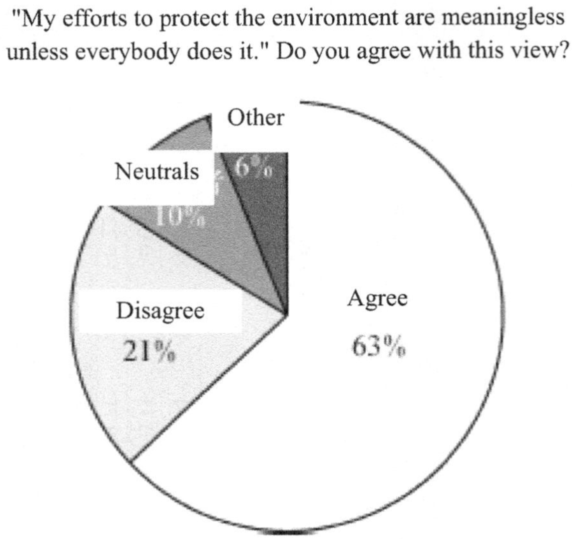

Fig. 6.5 Psychological analysis chart for Chinese residents' environmental protection. *Data source* Analyzed according to the data of CGSS2013

chose "enterprises"; 11.99% were unable to choose; only 6.64% held individual citizens responsible. Obviously, in the eyes of ordinary people, the government should be responsible for solving public problems. This reflected the "free-rider" problem in economics.

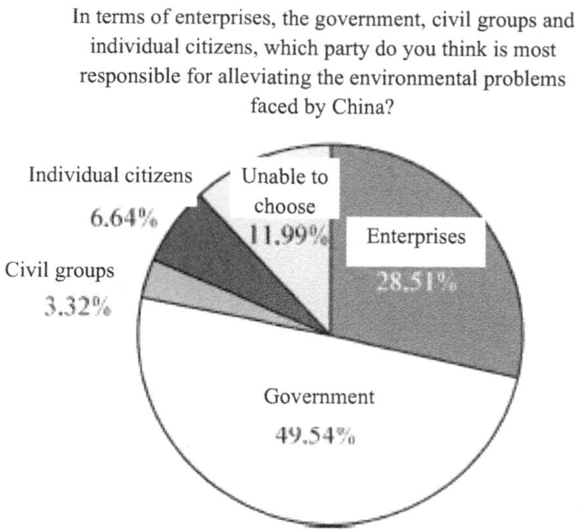

Fig. 6.6 Residents' subjective understanding: responsible party for China's environmental problems. *Data source* Analyzed according to the data of CGSS2013

6.2.2 *Chinese Residents Weak in Knowledge of Environmental Protection*

In the survey questionnaire, we set the following questions for participants to answer so as to get an understanding of residents' environmental awareness. See Figs. 6.7 and 6.8 for the questions and response category proportions. One of the questions is aimed at understanding residents' perception of the causes of environmental problems. Only 31% of the residents chose "Understand", and 49% chose "Do not understand".

As regards the solutions to environmental problems, 62.45% of the respondents chose "Do not understand", and only 15.52% answered "Understand". Obviously, environmental policies were not widely known to the masses, and needed more publicity.

In 2010 and 2013, CGSS surveyed residents' mastery of environmental protection knowledge, and raised the following 10 questions. This book compares the respondents' answers of the two years. See Table 6.7 for details.

Upon an analysis on related data of CGSS, we found that on the whole, Chinese residents did not have high environmental awareness, mastered little knowledge on environmental protection, and needed further popularization of such knowledge.

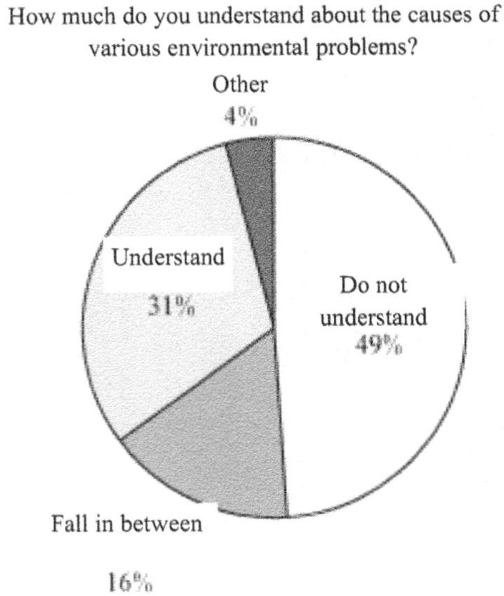

Fig. 6.7 Residents' subjective understanding: causes of environmental problems. *Data source* Analyzed according to the data of CGSS2013

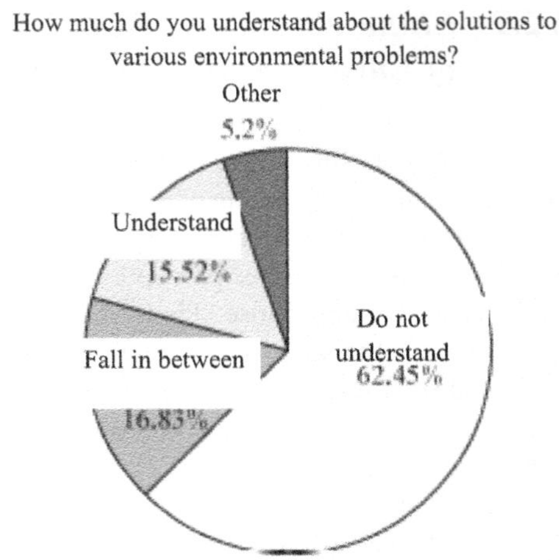

Fig. 6.8 Residents' subjective understanding: solutions to environmental problems. *Data source* Analyzed according to the data of CGSS2013

Table 6.7 Residents' mastery of environmental protection knowledge

Question description	Percentage of respondents answering "correct"		Percentage of respondents answering "incorrect"		Percentage of respondents "unable to choose"	
	2010	2013	2010	2013	2010	2013
Automobile exhaust will not threaten human health	12.34	12.38	80.99	75.46	5.43	11.1
The excessive use of chemical fertilizers and pesticides will lead to environmental damages	83.42	79.32	9.53	8.15	5.78	12.48
The use of phosphorous laundry detergent (powder) will not cause water pollution	12.83	11.4	61.79	54.84	24.81	32.69
The emission of fluorine from fluorine-containing refrigerators will be a factor destroying the atmospheric ozone layer	51.36	43.92	9.64	4.81	38.32	50.18
Acid rain is irrelevant to coal firing	10.70	8.53	43.85	39.61	44.34	51.77
Species are interdependent, and the disappearance of a species will produce a ripple effect	51.88	45.92	4.77	4.73	41.31	48.21
In an air quality report, Level 3 air quality is better than Level 1	10.87	8.61	24.98	21.44	62.06	69.86
Single-variety woods are more prone to plant diseases and insect pests	43.95	41.7	9.18	5.99	44.86	51.18
In a water pollution report, Class V (5) water quality is better than Class I (1)	7.82	5.31	15.23	13.95	74.75	79.64
The increase of carbon dioxide in air will be a factor causing climate warming	52.72	49.28	4.90	4.35	41.53	45.26

Data source According to the answers to questions l2401–l2410 in CGSS2010 and questions b2501–b2510 in CGSS2013
Unit %

Chapter 7
Empirical Analysis on Supply of Environmental Public Goods—A Case of Clean Air

Based on the foregoing theoretical model, this chapter takes China's pollution control as an object of study to probe into the feasibility of residents participating in environmental governance as the suppliers of environmental public goods, i.e. measuring residents' WTP for environmental protection.

For the empirical analysis on the supply of environmental public goods, we must consider China's special circumstances: the household registration system basically eliminates the possibility for residents to "vote with their feet" across provinces and cities; the separation of domicile and residence arising from population migration in the urbanization process may also severely affect relevant residents' WTP for environmental protection, or in other words, because of unclear home locations, relevant residents are unsure about their WTP for the environment where they live. But meanwhile, China's high housing price and residents' preference for property housing may cause resident identity solidification and regional space solidification. Owning property housing in a place is a deliberate choice made by residents. We can suppose that homeowners have higher requirements for environmental quality, and a stronger desire to see environmental quality improvement in their property location. This not only has a connection with their health, but also may bring the wealth effect of property appreciation. Based on the above analysis, this book puts forward the following assumptions.

Assumption 1 Homeowners have a higher WTP for environmental protection

Group difference is a matter of concern in China. According to the National Bureau of Statistics, China's Gini coefficient has exceeded the international warning line of 0.4 in recent years, and wealth disparity has become an accomplished fact. What is the difference in WTP for environmental protection between high and low income groups? Some scholars put forward that higher income spurred residents to pursue higher quality of life, and have a higher WTP for pollution control [159].

Additionally, environmental protection belongs to typical public goods with intergenerational positive externality. A family with or without children, and the number

J. Yang, *Effective Supply Mechanism and Path of Environmental Public Goods in China*,
Public Economy and Urban Governance in China,
https://doi.org/10.1007/978-981-99-2939-9_7

of children raised both directly influence residents' WTP for environmental protection. Another study shows that the degree of residents' concern for environmental quality, the differences in education among residents and their health conditions directly influence local environmental protection. Based on the above research, Assumption 2 is hereby presented.

Assumption 2 Residents' WTP for environmental protection varies among different income groups, different families (with children or no children), and individuals with different educational attainments.

The remaining part of this chapter will test the above assumptions.

7.1 Basic Modeling

The micro-data used in this chapter came from the data of CGSS2010, from which we obtained respondents' WTP for environmental protection, household registration places and real estate registration status. Besides, information with respect to individual income, household income, respondents' educational attainment, having children or not, and the like can be obtained from the database.

7.1.1 Data Source and Variable Description

1. Residents' subjective WTP for environmental protection

Residents' subjective WTP is the explained variable of the model. In survey statistics, there are two indicators which can reflect residents' WTP: first, "To what extent are you willing to pay higher prices for environmental protection"; second, "To what extent are you willing to pay higher taxes for environmental protection". This book brings the above two indicators into models, uses two models for respective regression, which is more persuasive. For sample processing, we denote "strongly willing" and "willing" by 1, and other answers by 0.

2. Home owning status

Home owning status is one of the primary explanatory variables herein. We choose it as an explanatory variable, because housing property is the best embodiment of residents "voting with their feet". The CGSS2010 had two surveys with regard to residential housing: first, the number of homes owned by a family; through which, we could understand whether homeowners have a higher WTP compared to non-homeowners; second, gross internal floor area. Through this indicator, we could know the effect extent of building area on residents' WTP. In the basic model regression analysis, this book applies the first indicator "owning a home or not", and introduces the second indicator "gross internal floor area" as a control variable into the model regression analysis. For variable processing, in consideration of biases and inaccuracies, the abnormal circumstance of gross internal floor areas below 10 m^2 and above 900 m^2 have been eliminated.

3. Control variables

To avoid biased or inaccurate estimations arising from omitted variables to the greatest extent, we consider controlling the following variables.

First, income status. Residents' income status not only directly influences their WTP for environmental protection, but also influences their home owning status. It is worth exploring that the CGSS questionnaire surveys respondents' household income and individual income, both of which are highly correlated to their WTP. But considering that residents would certainly start from the perspective of family burden when answering questions about WTP for environmental protection, the variable—household income is employed in the model herein, and then the logarithm thereof is calculated. According to the survey manual, "Respondents' annual household income includes wages, various bonuses, subsidies, dividends, stock dividends, insurance, pension, net operating income, bank interests, and gifts. In case of a rural laborer, his/her agricultural income, household and family conditions should be inquired about".

Second, educational attainment. Respondents' educational attainment directly influences residents' environmental awareness and behavior, as well as their household registration. For example, highly-educated personnel can be introduced as talents. In this book, the number of years of education received by respondents is used a variable: "uneducated" denoted by 0, "primary school" denoted by 6, "junior high school" denoted by 9, "senior high, technical secondary or vocational school" denoted by 12, "junior college" denoted by 15, "undergraduate" denoted by 16, and "postgraduate" denoted by 19, the unit of which is "year".

Third, health condition. Individual health condition is not only affected by local environmental pollution, but also closely related to individual WTP for environmental protection [21]. Thus, "self-rated health" as the proxy variable of individual health is brought into the model herein. CGSS requires respondents to assess their own health degree. We assign a score of 1–5 to the following levels, i.e. "very unhealthy" corresponding to 1, "relatively unhealthy" corresponding to 2, "fair" corresponding to 3, "relatively healthy" corresponding to 4, "very healthy" corresponding to 5, so as to obtain an ordinal selection variable for measuring self-rated health degree.

Fourth, other individual characteristics, including: (1) gender, "male" denoted by 1, "female" denoted by 0; (2) having offspring or not, "having children" denoted by 1, "no children" denoted by 0; (3) living condition, "urban residents" denoted by 1, "rural residents" denoted by 0; (4) marital status, including "married or cohabitating" denoted by 1, "single" and "divorced, separated or widowed" denoted by 0; (5) ethnicity, "minority" denoted by 1, "Han nationality" denoted by 0.

Fifth, air pollution indicator. This book analyzes air pollution that is the biggest concern of residents, and chooses NO_2 as an indicator measuring air pollution, mainly because compared to PM_{10} and $PM_{2.5}$, NO_2 is more likely to suspend in the air for a long time, and more harmful to human health [19].

Table 7.1 Descriptive statistics of variables

Variable	Unit	Observed value	Mean	Standard deviation	Minimum	Maximum
WTP for environmental protection		3656	0.425	0.494	0	1
Property housing	No	11,698	1.098	0.562	0	1
Annual household income	RMB	10,333	41,345	84,772	1	3,000,000
Number of years of education received		11,666	8.678	4.519	0	19
Number of children	No	11,669	1.771	1.344	0	11
Living condition		11,768	0.487	0.500	0	1
Marital status		10,642	0.886	0.318	0	1
Gender		11,783	0.482	0.500	0	1
Self-rated health		11,768	3.615	1.115	1	5
Age		11,780	47.30	14.68	17	96
Subjective happiness		11,767	3.766	0.883	1	5
NO_2	mg/m^3	11,042	0.043	0.010	0.015	0.067

The descriptive statistics of variables are shown in Table 7.1.

7.1.2 *Model Setting and Results*

According to the construction of the above variables, this book sets the following econometric model:

$$SWTP = \alpha h + X\gamma + \varepsilon \tag{7.1}$$

In Eq. (7.1), SWTP (Subjective Willingness to Pay) refers to residents' SWTP for environmental protection; h denotes residents and their spouses' home owning status; X denotes various control variables, e.g. household income, health condition, number of years of education received by respondents, gender, age, urban or rural household registration, and having children or not. α and γ are parameters to be estimated; ε denotes random error.

Table 7.2 Estimated results of basic models[1]

Explained variable	Model 1	Model 2	Model 3	Model 4	Model 5	Model 6	Model 7
	Residents' subjective WTP for environmental protection					Homeowners' subjective WTP for environmental protection	
Home owning	0.071*** (0.015)	0.053*** (0.016)	0.066*** (0.016)	0.051*** (0.017)	0.047*** (0.015)	0.085*** (0.021)	0.081*** (0.021)
Log of household income		0.0257*** (0.007)		0.0239*** (0.008)	0.022***(0.008)	0.026*** (0.008)	0.023*** (0.008)
Health condition		– 0.006 (0.008)		– 0.010 (0.009)	– 0.010 (0.009)	– 0.014 (0.010)	– 0.014 (0.009)
Number of years of education received		0.015*** (0.002)		0.014*** (0.003)	0.014*** (0.003)	0.014*** (0.009)	0.014*** (0.003)
Gender			0.037** (0.018)	0.034** (0.020)	0.030 (0.020)	0.034* (0.020)	0.032 (0.021)
Number of children			0.024*** (0.009)	0.033*** (0.010)	0.032*** (0.010)	0.031*** (0.010)	0.030*** (0.010)
Living condition			0.118*** (0.019)	0.054*** (0.023)	0.055** (0.024)	0.036 (0.024)	0.039 (0.024)
Marital status			0.027 (0.030)	0.009 (0.033)	0.008 (0.031)	0.013 (0.034)	0.014 (0.033)
Age			0.0003* (0.004)	0.001 0.004	0.002 (0.004)	0.0003 (0.004)	0.001 (0.004)
Age * age			– 0.004 (0.004)	– 0.004 (0.004)	– 0.004 (0.004)	– 0.002 (0.004)	– 0.002 (0.004)
NO_2	Control	Control	Control	Control	Control	Control	Control
Observed value	3389	2965	3042	2697	2662	2489	2460
R^2	0.007	0.035	0.043	0.053	0.053	0.056	0.055

Note: ∗ , ∗ ∗ and ∗ ∗ ∗ respectively stand for being significant at the levels of 10%, 5% and 1%; the number in brackets denotes standard deviation

This book employs relevant data of CGSS2010 for OLS regression. The results are described in Table 7.2.

Models 1–5 show the full-sample regression for the respondents. It could be observed that no matter any control variable or individual characteristic variable is added, homeowners' WTP for environmental protection is significantly higher than that of non-homeowners. In Model 1, no control variable or individual characteristic

[1] In the case that the "price residents are willing to pay for environmental protection" is substituted with the "tax residents are willing to pay for environmental protection", similar model simulation results can be obtained. Due to limited space, estimated results are omitted here.

variable is added, and the results thereof still suggest that homeowners have a strong WTP for environmental protection. In Model 3, after an individual characteristic variable is added, the coefficient slightly declines, but still reflects that homeowners have a high WTP. In Model 2, a control variable is added, but the results are still significant. In Model 4, all control variables and characteristic variables are added on the basis of the foregoing, fully demonstrating that homeowners' WTP for environmental protection rises by 4.11% compared to that of non-homeowners. Models 6 and 7 are empirical validations with respect to homeowners. The results suggest that homeowners' WTP for environmental protection greatly rises compared to that of the full sample, rising from 0.047 to 0.081. In case the living floor area is not controlled, homeowners' WTP for environmental protection will rise by 8.50% for every additional home.

The results of the above models all suggest that owning a home or not will influence residents' subjective WTP for environmental protection, and their WTP will be enhanced with the increase in the number of homes, which verifies the aforesaid Assumption 1.

7.2 Analysis on Heterogeneity of Residents' WTP for Clean Air

7.2.1 Variable Selection and Basic Description

According to basic models, we found that in addition to housing property, there are other factors (income, offspring, educational attainment, individual characteristics, etc.) influencing residents' WTP for environmental protection. To test the effect extent of some important indicators on residents' WTP for environmental protection, such variables as income, residents having children or not, and educational attainment are selected for analysis in this study.

The above three factors are selected for analysis, mainly based on the following reasons. First, residents' household income level, which comes from Easterlin's research conclusion—residents' happiness is related to income, and enhanced with the increase of income. In this study, samples will be classified into two groups by average income, namely, low income and high income groups. Second, whether married families have children or not: on this basis, samples will be divided into two groups, i.e. families with children and families without children, with the chief aim to test whether residents will produce an effect on environmental protection for protecting their offspring. Third, educational attainment: in existing studies, some scholars pointed out that higher-educated people would have a stronger WTP for environmental protection. Therefore, the author chooses 12-year education (senior high school) as the benchmark, and classifies samples into senior high school education background above, and senior high school education background or below.

7.2.2 Model Regression Results

Through the comparative regression of three variables, we test the heterogeneity of residents' WTP for environmental protection. The regression results are described in Table 7.3. Model 8 shows a comparison between low income and high income groups. Obviously, home-owning high income residents have a stronger WTP for environmental protection while low income residents do not show a strong willingness. Upon testing, Model 9 confirms that respondents with offspring have a more significant WTP for environmental protection, which is not only statistically significant, but also numerically higher than that of those who have no children. Model 10 suggests that respondents with high and low educational attainment can both pass the testing, but residents who have received senior high school education above have a higher WTP for environmental protection.

It follows that homeowners have a higher WTP for environmental protection, which appears more significant among high earners and married residents having children; high income families have a higher WTP than low income families, and married residents having children have a significantly higher WTP for environmental protection; higher educated residents are more willing to contribute to environmental protection. These conclusions fully demonstrate that residents integrate altruism and egoism. From the perspective of egoism, homeowners are more willing to pay for environmental protection, which is not only beneficial to their health, but also capable of increasing their home value. In addition, from the perspective of altruism, families with children are significantly more willing to pay for environmental protection than

Table 7.3 Heterogeneity of residents' WTP for environmental protection

Explained variable	Residents' subjective WTP for environmental protection					
	Model 8		Model 9		Model 10	
	Lower income	Higher income	Having no children	Having children	Lower educated	Higher educated
Housing property	0.028 (0.024)	0.062^{**} (0.028)	0.021 (0.056)	0.038^{**} (0.018)	0.048^{**} (0.020)	0.063^{**} (0.037)
Self-rated health	Yes	Yes	Yes	Yes	Yes	Yes
Air pollution level	Yes	Yes	Yes	Yes	Yes	Yes
Other control variables	Yes	Yes	Yes	Yes	Yes	Yes
Observed value	1957	705	235	2608	2327	361
R^2	0.043	0.040	0.081	0.048	0.035	0.047
F value	4.82	1.93	1.49	10.09	5.02	1.74

Note: $*$, $**$ and $***$ respectively stand for being significant at the levels of 10%, 5% and 1%; the number in brackets denotes standard deviation

those without children. This can be interpreted from the perspective of parental care for children, namely investment in human capital.

7.3 Residents' WTP for Environmental Protection Measured by the Happiness Measurement Method

7.3.1 Variable Selection of Happiness Measurement

The basic model suggests that residents have a significant WTP for environmental protection, and the stated preference method is mainly used in the basic model. However, some scholars considered that this method required respondents to directly give economic assessments by positing an environmental change, and thus might result in unreliable conclusions. To overcome the problem arising from the above method, a new public goods valuation method, namely happiness measurement method, is used to measure residents' WTP herein.

This book classifies the respondents into homeowners and non-homeowners to respectively measure and calculate their WTP for environmental protection. Environmental protection covers lots of items. Here we choose residents' highest concern: air pollution. The main reason for choosing this item is that it ranks first in terms of the degree of' concern expressed by residents and their subjective assessment on pollution severity in the questionnaires of CGSS2010 and CGSS2013 (See Tables 6.2 and 6.3). Additionally, according to the respondents' answers, the top four serious environmental problems include air pollution (41.82% of residents considered it very serious), desertification (39.58%), household waste pollution (39.29%), and water pollution (37.62%) (Fig. 6.1). It could be observed that air pollution, water pollution and household waste pollution, which are closely connected with the ordinary people's livelihoods and even lives, have become the biggest concern for residents.

Air pollution indicators include NOx, SO_2 and PM. According to the analysis conducted by Yang Jidong et al., this book chooses NO_2 as the indicator measuring air pollution, mainly because compared to PM_{10} and $PM_{2.5}$, NO_2 is more likely to suspend in the air for a long time, and more harmful to human health. The following regression equation is built.

$$H_{ic} = \beta P_c + \theta Y_{ic} + Z'_{ic}\eta + \mu_c + \varphi \tag{7.2}$$

In Eq. (7.2), the subscript i denotes the respondents; the subscript c denotes the province where the respondents live; H denotes residents' subjective happiness; P denotes regional pollution, the indicator of which is still NO_2; Z' denotes the control variable, including health condition, number of years of education received and other individual characteristics (e.g. children, gender, and urban residents); μ denotes the regional control variable; φ denotes random error.

As regards the data used for happiness measurement, apart from the air pollution indicator that comes from the website of the National Bureau of Statistics, the remainder comes from CGSS2010.

According to Eq. (7.2), we can estimate that for every 1 unit of pollution decrease, residents' happiness will increase by $\widehat{\beta}$, i.e. the marginal utility of pollution; similarly, residents' happiness will increase by $\widehat{\theta}$ for every income increase of RMB 1. The perfect differential of the above function is set as $\Delta H = 0$, Then, we can obtain the marginal rate of substitution between environmental pollution and income level, i.e. the portion of income (yield) Y that residents are willing to pay for every 1 unit of pollution P decrease. $MRS_{PY} = -\frac{\Delta Y}{\Delta P} = \frac{\frac{\partial H}{\partial P}}{\frac{\partial H}{\partial Y}} = -\frac{\hat{\beta}}{\hat{\theta}}$, we assume that the happiness decision function is the logarithm of income. Then, Eq. (7.2) can be translated into Eq. (7.3).

$$H_{ic} = \beta P_c + \theta \ln Y_{ic} + Z'_{ic}\eta + \mu_c + \varphi \tag{7.3}$$

Upon derivation by the above method, we can obtain:

$$AWTP = \frac{MU_P}{MU_Y} = -\frac{\frac{\partial H}{\partial P}}{\frac{\partial H}{\partial Y}} = -\frac{\widehat{\beta}}{\widehat{\theta}} \cdot \overline{Y} \tag{7.4}$$

In Eq. (7.4), $\overline{Y}$ denotes the average income of the group; MU_P denotes the utility from pollution; MU_Y denotes the marginal utility from income.

Thus, the WTP obtained is the AWTP (Average Willingness to Pay). In consideration of the crowding-out effect of China's high housing price on residents' consumption, residents owning more than 1 home are additionally selected into homeowners in this part of regression analysis.

7.3.2 Model Regression and Conclusion

1. Effect of housing property on residents WTP for environmental protection

According to Eq. (7.3), a regression analysis of the effect of housing property on residents' WTP for environmental protection is carried out, and the results are described in Table 7.4. Models 11, 12 and 13 stand for non-homeowners; Models 14, 15 and 16 stand for residents owning one home; Models 17, 18 and 19 stands for residents owning more than one home. For each type of residents, we conduct the regression without control variables, the regression with controlled individual characteristics, and the regression with controlled educational attainment respectively. The conclusion suggests that residents owning more homes have a higher WTP for environmental protection, which is numerically greater than that of those owning only one home. Their WTP is more than 10 times that of non-homeowners. Seen from the percentage of residents' WTP to household income, upon the addition of the control variable, none-homeowners' WTP for environmental protection is less than 1% of their household income. We can suppose that these residents are saving money for buying a home, or still in the stage of enough food and clothing, with no time for other matters. The WTP of residents owning one home also accounts for less than 1%. We can suppose that affected by repayment pressure, they are subjectively less willing to make a contribution to environmental protection. However, when residents own more than one home, their WTP for environmental protection is higher than that of the former two (non-homeowners and homeowners owning only one home) in terms of the absolute value of money paid and the percentage of WTP to household income. This numerically verifies the assumption that homeowners have a stronger WTP for environmental protection.

Obviously, the results obtained by use of the happiness measurement method are identical with those obtained by the descriptive revelation method. Upon the addition of individual characteristics and control variables, homeowners' WTP for environmental protection is higher than that of non-homeowners, and residents owning more homes have a higher WTP.

2. Analysis on heterogeneity of WTP of residents owning different housing properties

Next, we further employ the happiness measurement method to test the heterogeneity of residents' WTP. Non-homeowners, residents owning one home, and residents owning more than one home are grouped according to income and offspring, and the test results are described in Table 7.5. In Model 20, higher income homeowners' WTP (numerically) for environmental protection is over twice that of non-homeowners; in Model 22, highly-educated residents' WTP (numerically) is approximately 10 times that of less-educated residents, but the credibility of this data needs to be further verified, because the observed value only covers 109 persons; as regards residents having offspring, Model 21 distinguishes between residents owning one home and those owning more than one home in terms of WTP for environmental protection, and the results suggest that the WTP of residents owning more than one home and children is roughly 8 times that of residents owning only one home.

It follows that in the heterogeneity regression test on residents' WTP, high income groups and residents who have received longer education still have a higher WTP for environmental protection. Considering the offspring factor, we have carried out a regression on the WTP of homeowners and residents owning more than one home, and found that among residents who have offspring, those owning over one home have a higher WTP. It should be particularly noted that high income residents' absolute WTP for environmental protection is far higher than that of low income residents, but seen from the percentage of WTP to household income, the WTP of low income families accounts for 1.53% while that of high income families only accounts for 0.67%.

3. Model conclusion

These basic models suggest that homeowners' WTP for environmental protection is significantly higher than that of non-homeowners. It appears fairly significant from

Table 7.4 Effect of housing property on residents' AWTP for clean air

Model	Object	Control variable	Annual average household income/RMB	β	θ	AWTP	Percentage of AWTP to household income/%	Sample size	R^2
11	Non-homeowners	No control	33,505	− 1.019 (3.471)	0.073*** (0.026)	465.24	1.389	733	0.022
12		Individual characteristic		− 0.694 (3.703)	0.125*** (0.032)	185.68	0.554	640	0.048
13		Educational attainment		− 0.996 (3.729)	0.120*** (0.033)	278.01	0.830	632	0.050
Model	Object	Control variable	Annual average household income/RMB	β	θ	AWTP	Percentage of AWTP to household income/%	Sample size	R^2
14	Homeowner	No control	41,983	− 1.283 (0.927)	0.145*** (0.007)	371.20	0.884	8984	0.044
15		Individual characteristic		− 0.988 (0.969)	0.129*** (0.017)	322.17	0.767	8221	0.058
16		Educational attainment		− 0.948 (0.091)	0.114*** (0.018)	348.28	0.830	8146	0.063
17	Residents owning more than one home	No control	82,725	− 2.306 (1.996)	0.077*** (0.016)	2480.68	2.999	1395	0.018
18		Individual characteristic		− 2.801 (2.145)	0.076*** (0.017)	3049.30	3.686	1239	0.025
19		Educational attainment		− 2.901 (2.157)	0.076*** (0.018)	3154.00	3.813	1229	0.031

Note: ∗ , ∗ ∗ and ∗ ∗ ∗ respectively stand for being significant at the levels of 10%, 5% and 1%; the number in brackets denotes standard deviation

Table 7.5 Different groups of residents' AWTP for clean air

Explained variable	Residents' subjective happiness					
	Model 20		Model 21		Model 22	
Family status	Homeowners		Residents owning one home	Residents owning more than one home	Homeowners	
	Lower income	Higher income	Having children	Having no children	Lower educated	Higher educated
NO_2	− 1.728 (1.175)	− 0.390 (1.367)	− 0.695 (0.978)	− 2.015 (2.167)	− 1.018 (1.013)	− 11.589 (8.381)
Log of household income	0.113^{***} (0.009)	0.0581^{**} (0.026)	0.108^{***} (0.008)	0.074^{***} (0.018)	0.127^{***} (0.008)	0.027^{***} (0.079)
Annual average household income/ RMB	18,961	100,960	39,497	74,749	36,670	90,154
Other control variable	Uncontrolled	Uncontrolled	Controlled	Controlled	Controlled	Controlled
Observed value	6 578	2 401	7 986	1 199	7 749	109
R^2	0.020	0.002	0.056	0.022	0.055	0.029
F value	67.61	2.43	58.32	4.01	55.41	2.26
WTP	290.21	677.82	253.34	2,040.61	295.07	38,131.19
Percentage of WTP to household income/%	1.53	0.67	0.64	2.73	0.80	42.30

Note: ∗ , ∗ ∗ and ∗ ∗ ∗ respectively stand for being significant at the levels of 10%, 5% and 1%; the number in brackets denotes standard deviation

Models 1–7, and the significance and direction of all explanatory variables remains consistent in all models. Thus, the estimated results are relatively stable.

The estimated results of control variables can bring use some enlightenment.

First, higher-educated respondents have a stronger WTP for environmental protection, which is significant at the level 1% in all models. This is consistent with the research conclusion drawn by Zheng et al. [159].

Second, families with children have a higher WTP for environmental protection. Previously, some scholars have drawn the following empirical study conclusions: Air pollution severely affects our physical health; SO_2 pollution is highly correlated with infant mortality; CO has the most significant impact on infant mortality. Thus, families with children are more concerned about environmental pollution [168,169].

Third, such characteristics as religions, party members, male and urban residents improve residents' WTP. It should be added that respondents' physical health is inversely proportional to their WTP for environmental protection, which is insignificant, but has consistent notations in all models. This indicates that less healthy groups are more concerned about environmental pollution.

At last, we could find that in case the specific AWTP of residents is measured by the happiness measurement method, the increase of income significantly enhances residents' happiness while pollution lowers their happiness. Besides, the data suggests that homeowners' AWTP is 1.25–1.75 times that of non-homeowners; the AWTP of residents owning more than one home is 5.3–15.4 times that of non-homeowners. Using the stated preference method, we could conclude that the WTP of high income groups is significantly higher than that of low income groups; but when using the happiness measurement method, we could observe that the absolute amount of high income families' WTP for clean air (homeowners are willing to pay RMB 677.82 for every 1 μg NO_2 decrease per cubic meter of air) is higher than that of low income families (RMB 290.21). In terms of household income, the percentage of the WTP of high income groups to household income (0.67%) is far lower than that of low income groups (1.53%).

Chapter 8
Case Study on Supply of Environmental Public Goods—"Five Water Co-governance" in Zhejiang

With the improvement of people's living standards and urbanization quality, many private demands have been translated into public demands. Centralized drainage, centralized travel, centralized waste disposal and other urban public demands are specific to the process of urbanization. In terms of water resources, many cities of China are faced with water environment deterioration, and all sorts of water pollution incidents reflect a series of problems, including insufficient urban sewage treatment capacity and incapability of quick digestion locally. The development of urban water pollution prevention and control relates directly to the improvement of human settlement quality and urban public health security. In recent years, with the increasing severity of urban water pollution, a discussion on the mechanism of water pollution prevention and control has been conducted in the academic circles. The "free-rider" problem of public goods can be extended to the conclusions that government governance is the most efficient, and the government may fail. Then, the market mechanism can be introduced into the supply of public goods, followed by a market-oriented and socialized multi-subject governance mode. Which operation mechanism is more effective for the urban water pollution prevention and control in China? The achievements made by Zhejiang in "Five Water Co-governance" have brought us some in-depth thoughts and a series of inspirations.

J. Yang, *Effective Supply Mechanism and Path of Environmental Public Goods in China*,
Public Economy and Urban Governance in China,
https://doi.org/10.1007/978-981-99-2939-9_8

8.1 Status and Prevention and Control Difficulty of Water Pollution in Zhejiang

8.1.1 Status and Governance Reason of Water Pollution in Zhejiang

"At sunrise riverside flowers redder than fire; in spring green waves grow as blue as sapphire". The verses reflect not only the unique beauty of nature in the lower reaches of the Yangtze River, but also the indispensable role of water in the beauty of nature. Water is a precious resource, without which humans cannot live. Water constrains socioeconomic development, especially of Zhejiang, which can be summarized into a combination of "seven portions of mountains, one portion of water and two portions of fields".[1] Zhejiang is a typical province that is geographically small and short of resources, with a plain area of only 22,000 square kilometers, and a per capita water availability of 1760 m^3, approaching the globally recognized warning line of 1700 m^3. As one of China's most populously and economically dense provinces, Zhejiang contributed 6% of the national GDP on 1% of the country's land with 4% of the national population. In the process of rapid industrialization and urbanization, Zhejiang's water environment has been polluted to different extents, and some of its rivers and lakes have been in the state of sub-health. In 2013, 27 of its province-controlled surface water sections were blow Grade V,[2] and 32.6% of sections failed to meet functional area requirements. Moreover, the eight river systems (eight major rivers, from north to south, including Tiaoxi, the Zhejiang Section of the Grand Canal, the Qiantang River, the Yong River, the Jiao River, the Ou River, the Feiyun River and Ao River) of Zhejiang have all been polluted to different extents.

8.1.2 Source Analysis on Water Pollution in Zhejiang

Considering such serious water pollution, governance is a matter of great urgency. In general, one of the causes of water pollution is the change in the laws of nature and the pollution of water sources by minerals in soil, commonly known as natural pollution; the other cause is man-made pollution, which is the main reason in our view. Man-made pollution refers to pollution caused by our life and production activities, often including industrial wastewater pollution, domestic sewage pollution, agricultural pollution, as well as water pollution arising from municipal solid waste.

---

[1] Zhejiang covers a land area of over 100,000 square kilometers, including mountains 70.4%, plains 23.2%, as well as rivers and lakes 6.4%. There is a saying that Zhejiang consists of "seven portions of mountains, one portion of water and two portions of fields".

[2] According to the environmental functions and conservation goals of surface water, China's surface water is classified into five grades by function: Grades I, II, III, IV and V. Grade V is mainly applicable to agricultural water areas and general landscape water areas; water below Grade V means that the water pollution exceeds Grade V.

According to research findings, the causes of the water environment pollution in Zhejiang mainly include agricultural pollution, domestic pollution, pollution from livestock and poultry, and industrial pollution resulting from low-level industries which accounts for the largest percentage. In 2013, in terms of the four heavy pollution industries concerning printing and dyeing, papermaking, tanning and chemical industries, their output value accounted for less than 37% of the province's gross industrial output value, but their COD and ammonia nitrogen emissions accounted for 67% and 80% of the province's industrial emissions; the output value of electroplating and tanning industries accounted for less than 5% of the province's gross industrial output value, but the total chromium emission thereof accounted for 92% of that of the province. Domestic pollution comprises household waste and domestic sewage pollution. Domestic sewage refers to the wastewater produced by the Party and government agencies, schools and residents in daily life, including household water discharge concerning toilet excrement and urine, laundry wastewater, bath water and kitchen wastewater, as well as water discharged from commercial areas, hospitals and playgrounds. With large consumption and complicated components, domestic water directly flows into water bodies without treatment, thereby causing severe water environment pollution. Municipal solid waste pollution mainly involves kitchen waste, plastic waste, waste paper, broken glass, and metal ware. With the rapid development of the economy, the output of residents' household waste is on the rise. In the stacking or landfill process, such waste produces a large quantity of acid and alkaline substances; the discharged wastewater containing mercury, lead and cadmium infiltrate surface water or underground water, and may result in black and smelly water, useless shallow groundwater, water quality deterioration, and other consequences.

As a developed area in China, Zhejiang is densely populated. With the agglomeration of population, the acceleration of urbanization and the improvement of people's living standards, it is faced with an increasing amount of sewage discharge.

Water environment pollution seriously affects social stability. Water environment problems could not only affect the improvement of people's livelihood, but also bring challenges to social stability. Around 2013, environmental complaints, especially complaints arising from water pollution, showed a significantly rising trend. Such incidents are likely to break through geographical restrictions, and produce ripple effects. In case of being mishandled, they could easily produce a negative impact on social stability. Since the causes have been found, how should water pollution be controlled?

8.2 "Five Water Co-governance" with Public Participation

In June 2013, Xia Baolong, then-Secretary of the CPC Zhejiang Provincial Committee, inspected Pujiang. He stressed the necessity to start with water governance, unswervingly promote transformation and upgrading, speed up efforts to implement the new development idea that "lucid waters and lush mountains are invaluable assets", and directly propel the introduction of the "Five Water Co-governance", No. 1 project of the CPC Zhejiang Provincial Committee and the provincial government. On November 29, 2013, the "Five Water Co-governance" was proposed at the Fourth Plenary Session of the 13th CPC Zhejiang Provincial Committee, including the following specific contents: sewage treatment, flood control, water-logging draining, water supply guarantee, and water saving.

At the end of 2013, the CPC Zhejiang Provincial Committee and the provincial government made a decision on and arrangements for the "Five Water Co-governance": We would rather temporarily abandon partial interests (at the cost of 1% economic growth every year) to start with water governance, drive industrial transformation and upgrading, and never bring mud and slush into moderate prosperity.

8.2.1 *Specific Meaning and Planning of "Five Water Co-governance"*

With the ambition for "reconstruction" and the resolute decision made by the CPC Zhejiang Provincial Committee and the provincial government, the leading officials of Zhejiang at all levels actively responded to the call, and fully sounded the bugle call to implement "sewage treatment, flood control, water-logging draining, water supply guarantee, and water saving", commonly known as "Five Water Co-governance". Zhejiang vividly likened the project to five fingers. As shown in Fig. 8.1, five fingers splaying out stand for different tasks, focusing on both overall planning and key points; five fingers bending into a fist imply starting with water governance to ensure the effective combination of transformation and upgrading.

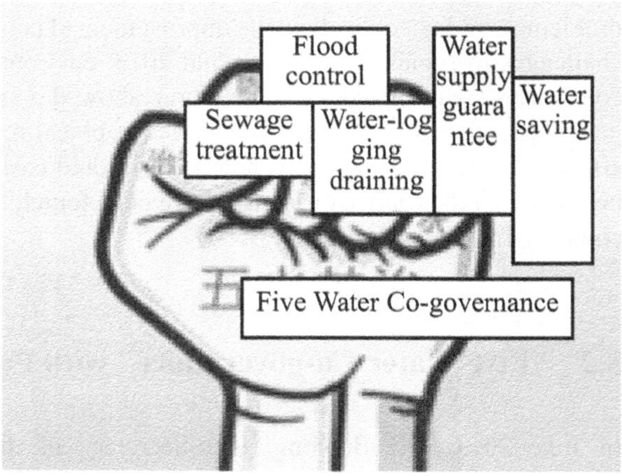

Fig. 8.1 Five-finger picture of "five water co-governance"

1. Sewage treatment

Sewage treatment is the "thumb", which is mainly manifested in urban domestic sewage treatment, rural domestic sewage treatment, industrial pollution treatment and agricultural pollution treatment. From the perspective of social response, the ordinary people have the most direct feeling and deepest abhorrence for sewage. From the perspective of practical operation, sewage treatment is most capable of driving the overall situation and the most effective. If sewage can be controlled, the ordinary people will give a thumbs-up. With the core of water quality improvement, sewage treatment includes the following contents: executing dredging, pollution interception and integrated river course renovation, enhancing security guarantee for drinking water sources, fully implementing the river chief system, and carrying out whole-basin water control. From 2013 to 2016, in the process of sewage treatment, Hangzhou closed down 6847 farms with substandard pollution control facilities, and completed the control acceptance of 789 farms with the raising quantity of above 50 (pigs and cattle). Live pig stocks of the city decreased from 2.1923 million as of the end of 2013 to 1.2691 million as of the end of 2016. Yuhang District closed down 9200 mu[3] of snakehead farming ponds and 3000 mu of soft-shelled turtle farming ponds. With the planting of lotus roots, these ponds turned into lotus ponds and seas of flowers. Besides, the output value of lotus sprout is greatly higher than that of fish farming, with significant industrial transformation benefits. Sewage treatment started from water control to make the masses give a thumbs-up.

2. Flood control

Popularly speaking, flood control is focused on solving the problem of harmony between man and floods (nature). This requires not only satisfying the people's needs for survival and development and safeguarding people's lives and property, but also meeting the need for the floods to "run into the sea". As regards water resources, humans are faced with three problems, namely "excessive water", "scarce water" and "foul water". Sewage treatment is aimed at solving "foul water", and flood control is aimed at solving "excessive water". In all parts of China, heavy rain and floods may block travel modes, turn roads into rivers, cause cars to be flooded, and even lead to house collapse and life-threatening injuries. Flood control mainly refers to flood prevention measures, strategies and methods. For residents in the coastal province Zhejiang, this project was a matter of great urgency. The main objective of flood control is to advance the construction of reservoir reinforcement, dyke strengthening, drainage expansion and other works, strengthen the overall planning for basins, and implement both dredging and blocking to control the fierce flood.

3. Water-logging draining

Similar to flood control, water-logging draining is also aimed at solving "excessive water". It mainly includes cutting through beheaded rivers, and opening up new river courses, with a focus on eliminating areas prone to flooding and water-logging. In

[3] 1 mu $\approx$ 666.67 square meters.

real life, torrential rain tests a city's municipal works. What are the causes of urban water-logging in a rainstorm? Are sewers are unobstructed? Is the water pipe layout reasonable? Is the facility maintenance in place in ordinary days? Does the early-warning system for rainstorm water-logging work? Focusing on flood prevention and stagnant water draining, Hangzhou carried out eight major tasks, including working out plans, performing work in a planned manner, strengthening the construction of flood control regulations, improving the early-warning system for rainstorm water-logging, upgrading the infrastructure detection system, strictly executing facility construction standards, elevating the maintenance level of drainage facilities, and boosting the capability of flood control and emergency rescue. The eight works include prevention, implementation and routine maintenance, present an overall plan for the stagnant water drainage works in Hangzhou, and play a great role in such works.

4. Water supply guarantee

Water supply guarantee means ensuring the conformance of the amount and quality of water supply to certain standard requirements. It is aimed at solving the water resource problem—"scarce water". Properly speaking, water supply guarantee is focused on advancing source opening-up, diversion, elevation and other works to guarantee the source of drinking water and improve the quality thereof. Water is the source of life. In a manner of speaking, we will be unable to live without water. Many parts of Zhejiang have implemented some works concerning drinking water purification renovation, and extension of feed water booster pump stations to ensure that residents could use high-quality water with ease (improving the past obstructed water supply).

5. Water saving

Opening up the source and regulating the flow, water saving includes refitting instruments, reducing leakage losses, collecting water for recycling, and rationally using water resources, with a focus on lowering water consumption. Water supply guarantee and water saving are mainly aimed at solving "scarce water".

Take Hangzhou as an example. Since 2013, Hangzhou has vigorously carried out the construction of county-wide water-saving society, and Yuhang District has been determined as one of the first batch of water-saving society construction counties (cities/districts) by the provincial government and passed the national and provincial acceptance. As of 2017, Hangzhou successfully established 80 water-saving enterprises, with the establishment rate of water-saving enterprises in five high water consumption industries up to 36%; established 138 provincial water-saving housing estates, with the coverage of water-saving residential estates up to 17.03%; established 66 water-saving agencies, with the coverage of municipal water-saving agencies up to 69.5%; established 8 water-saving irrigation areas. As regards enterprises with high water consumption and highly polluting enterprises, Hangzhou took the measures of closing, suspension, merging, shifting and relocation. Especially in 2014, it completed the renovation of 479 major industry enterprises, and eliminated 522 enterprises involved in outdated production facilities.

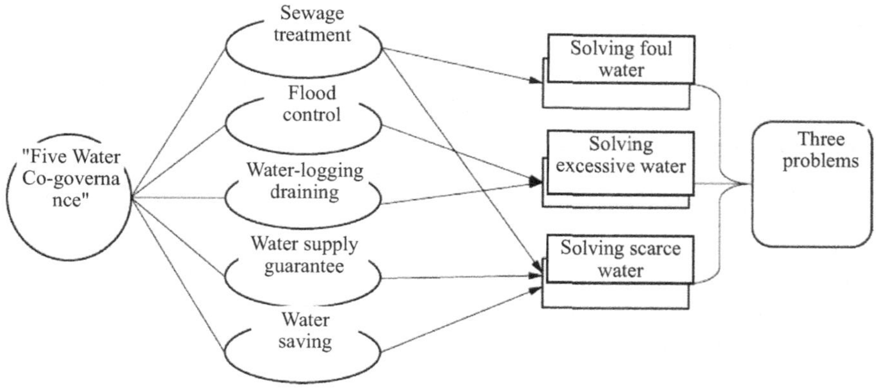

Fig. 8.2 "Five water co-governance" solving three water resource problems

It follows that the "Five Water Co-governance" in Zhejiang was carried out with a focus on three water utilization problems, namely, "excessive water, scarce water and foul water". As shown in Fig. 8.2, flood control and water-logging draining are mainly aimed at solving "excessive water"; water supply guarantee and water saving are mainly aimed at solving "scarce water"; sewage treatment is aimed at solving not only "foul water", but also "scarce water" to some extent. Focusing on solving the three problems regarding water resources, the "Five Water Co-governance" is objective-oriented.

8.2.2 Mechanism Design for "Five Water Co-governance"

1. Leading officials attaching importance ideologically and participating practically

In early 2014, the CPC Zhejiang Provincial Committee and the provincial government officially established a leading group for "Five Water Co-governance", where the secretary of the provincial committee and the governor served as double group leaders, and six deputy provincial leaders served as deputy group leaders to make overall plans for and coordinate the water control work. The province transferred more than 40 backbone officials from the Leading Group Office to implement centralized office work and solid operation. The office set a water control schedule. In accordance with the requirements of the schedule (solving prominent problems within three years (2014–2016), with significant results; basically solving all problems within five years (2014–2018), with all-round changes; ensuring no problems basically within seven years (2014–2020), realizing qualitative changes, and never bringing mud and slush into moderate prosperity) and the roadmap of "Five Water Co-governance, Starting with Sewage Treatment", starting with the poorest river in the province, Zhejiang took the lead in launching the integrated treatment of water environment in the Puyang River, rapidly spread it throughout the province, made breakthroughs one by one and deepened it in stages while advancing in an orderly fashion.

From planning to concrete actions, the leading officials of Zhejiang not only made a call for water governance, but also put it into practice, and led by example. This is fairly critical in the governance of public goods.

2. Establishment and improvement of system construction

To ensure the advancement of the "Five Water Co-governance" pursuant to the action plan, in addition to the leading role played by leading officials, operation guarantee mechanisms should be equipped, which could realize "guidable planning, normal project progress, fund support, in-place supervision, guidable assessment, sci-tech support, binding regulations, and unified commands".

The organizational guarantee mechanism: This requires building a three-level (province-municipality-district/country/city) leading group mechanism for "Five Water Co-governance", which includes offices, composed of backbone officials designated by the organization department, with the persons chiefly in charge of environmental protection, housing construction, and city management serving as the office directors.

The problem solving mechanism: This requires solving problems within a limited time, leaving traces in the system for ease of query and supervision, and performing peer-to-peer execution in solving problems to avoid repetitions and procrastination, and to ensure smooth settlement.

The accountability mechanism: A set of policies characterized by "one river with one policy" have been developed, and a mayor supervision sheet has been prepared. This requires taking measures in a timely manner in case of identifying any problems, giving notices and carrying out negotiations for severe ones, and deciding whether to submit such problems to the mayors for supervision based on the nature thereof.

The linkage mechanism: During water governance, one of the major problems is zone division. For instance, even when the lower reaches are rectified, the sewage discharged from the upper reaches may still pollute rivers. In the water governance process, Zhejiang established a mechanism for inter-basin coordinated governance of upper and lower reaches, interregional coordinated governance, overseas linkage and interdepartmental linkage.

The early-warning mechanism: For the water quality monitoring of river courses, the tricolor (orange, yellow and red) early-warning mechanism has been launched. The water governance offices of all districts should give notifications and send an early warning for over-limit river courses pursuant to the early-warning standards. For example, in the treatment process of the Fuyang River in Hangzhou, a red alert would be sent in case that the river course monitoring results exceeded the standard for black and smelly rivers defined in the *Clean-up and Acceptance Standards for Garbage Rivers and Black and Smelly Rivers in Zhejiang Province*. For river courses included in the red alert list, river chief units, as well as towns/townships (sub-districts) to which such river courses belonged must fulfill rectifications within one week upon receipt of an alert notification, ensure that all indicators could reach the standard for getting rid of black and smelly rivers, and submit the check and rectification conditions to the district water governance office in writing. A yellow alert would be sent in case that the river course monitoring results exceeded the Grade V standard defined in the *Environmental Quality Standard for Surface Water*, or the standard for black and smelly rivers defined in the *Guidelines for Rectification of Black and Smelly Water in Cities* (J.C. No. 130 [2015]). For river courses included in the yellow alert list, river chief units, as well as towns/townships (sub-districts) to which such river courses belonged must fulfill rectifications within half a month upon receipt of an alert notification, ensure that all indicators could reach the standard for Grade V water in the *Environmental Quality Standard for Surface Water*, and the standard defined in the *Guidelines for Rectification of Black and Smelly Water in Cities*, and submit the check and rectification conditions to the district water governance office in writing. An orange alert would be sent in case that river course monitoring results exceeded Grade IV standard defined in the *Environmental Quality Standard for Surface Water*. For river courses included in the orange alert list, river chief units, as well as towns/townships (sub-districts) to which such river courses belonged must attach great importance after receiving an alert notification, strengthen management and control, improve the long-term management mechanism, and strive to improve water quality.

The assessment supervision mechanism: This requires implementing a special assessment mechanism, carrying out special supervision, and performing media interaction supervision.

3. Mobilization of public participation

Water pollution is also affected by individual misconduct. Business owners' improper conduct could cause industrial pollution; peasants' improper agricultural production could lead to agricultural pollution; residents' improper lifestyle could result in domestic pollution… The fundamentality of water governance lies in people governance. If human behavior cannot be effectively constrained, there will be no end to pollution control. Only when the people's misconduct is curbed can pollution be prevented or reduced.

For the purpose of responding to the "Five Water Co-governance" launched by the CPC Zhejiang Provincial Committee and the provincial government, and motivating the public to actively participate in river regulation, the Green Zhejiang "Water Co-governance" campaign was jointly initiated by the non-governmental environmental protection organization Green Zhejiang and its Party branch—Rainbow Life Party Branch of Wenhui Sub-district, Xiacheng District, Hangzhou, in cooperation with water governance experts among the members. They took the opportunity of the second phase of Party's mass line education practice activities to go deep into grass-roots units, and to launch this public service campaign, with the theme of "Our Water: the Water in Our Homeland; Our Responsibility: Five Water Co-governance", which is characterized by public participation in water governance under the leadership of Party members in support of the "Five Water Co-governance" in all parts of the province. This campaign includes three concrete moves, i.e. convening the interested parties' round-table conference, propelling the building of the non-governmental

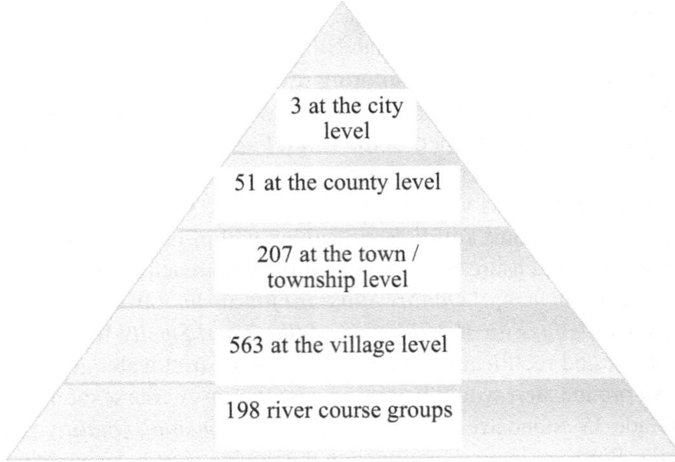

Fig. 8.3 Water governance network contact groups at all levels in Hangzhou's water governance

river chief system, and publicly collecting water governance proposals. Hangzhou has built more than 1100 WeChat groups and QQ groups concerning water governance, as shown in Fig. 8.3, forming a pyramidal network for quick contact, handling and feedback with respect to relevant problems. Besides, it has developed an APP regarding Hangzhou's river water quality to publish water quality data in a timely manner. This platform allows the public to directly dial a river chief's number, make one-touch complaints, and learn about the river chief's river patrol trajectory, performance of duty, etc. for ease of subsequent supervision and accountability.

Gaining beauty and prosperity from water, the "Earthly Paradise" Hangzhou is saturated with the charm of lower reaches of the Yangtze River, and contains originality across generations. Zhejiang issued a series of policies, including the *Implementation Plan of Zhejiang Province for Beautiful River and Lake Construction (2018—2022)*, and the *Measures for the Administration of "Beautiful Rivers and Lakes" Evaluation in Hangzhou (for Trial Implementation)*, and actively mobilized the general public to participate in the governance of river and lake water. Figure 8.4 shows the public participation status in the process of Hangzhou's water governance, which adopted the river (lake) chief system as the main means under the leadership of the government.

Building a social participation network composed of 1418 nongovernmental river chiefs, 7512 river patrol volunteers, and 5800 river cleaners (part-time information officers)

Hangzhou set up 5847 river chiefs and 1119 lake chiefs at all levels. In the appraisal of 2018, 39 river courses including Fenglin Port were appraised as municipal "Beautiful Rivers and Lakes", among which, four river courses were appraised as provincial "Beautiful River Courses". 1455 nongovernmental river chiefs made silent contributions to the beauty. In 2018, the city carried out river (lake) chief training for

Fig. 8.4 Public participation in Hangzhou's water governance

261 times, with more than 14,500 person-times of trainees, who realized the transformation from "Beginners" to "Experts" in the process of water governance. In addition, Hangzhou organized the full-coverage supervision of the two-level (city and county levels) river chief system, actively expanded channels for public participation, improved the river chief system and the voluntary supervisor system, fully introduced media supervision, and found 912 problems through self-inspection and 163 problems through municipal supervision and inspection, which were all rectified.

8.3 Enlightenment from the "Five Water Co-governance" in Zhejiang

8.3.1 Government-Led Orderly Participation of the Media, Social Organizations and the Public

The effective water governance in Zhejiang was a result of the concerted efforts made by residents (including volunteers), social organizations, the government, the media, and so on. Businessman Jin Zengmin's "offer of a reward to the environmental chief for a river swim" was a trigger, but if he did not have received positive response and support from all walks of life, "our water" would still have been "sewage". The ordinary people (setting off an upsurge in offering a reward for swimming) who had the deepest feelings, the environmental chief in the teeth of the storm, the chairman of the Standing Committee of the Zhejiang Provincial People's Congress, the governor and the secretary of the provincial committee, as well as the subsequent news media coverage all greatly facilitated the positive action taken by the governments of Zhejiang at all levels and the residents' cooperation. The government guided and strengthened public opinion publicity, and mobilized officials, the masses, entrepreneurs, Zhejiang businessmen outside the province, ethnic Chinese and overseas Chinese to volunteer to make donations or work for their offspring and their own lucid waters and lush mountains. In terms of mobilizing the masses, on the basis of fully considering the Chinese ordinary people's customs, and grasping residents' face-saving and enterprising mentality, Zhejiang introduced village regulations and nongovernmental agreements on "Five Water Co-governance" at all parts of the province, and implemented the "three doorstep assignments" responsibility document, etc. to jointly advance the institutionalization and socialization of river cleaning.

8.3.2 Elevation of Public Goods Governance to a Position of Equal Importance with Food and Public Security

One important reason for Zhejiang's achievements in the "Five Water Co-governance" is that the government sector not only faced up to the existing problem, but also elevated it to a position of equal importance with food security. From the perspective of governance, Zhejiang and its cities and counties all built a working mechanism for "Five Water Co-governance", where the top Party and government leaders served as the leading group leaders, and the Party committee or government leaders with specific duties served as directors of water governance offices; and transferred some personnel for centralized office work. The province transferred a total of nearly 1000 comrades to water governance agencies at all levels. All parts of the province even established an operational command office for "Five Water Co-governance" to implement centralized command and operations as scheduled; some cities and counties even built an information platform for "Five Water Co-governance" to implement real-time monitoring and command.

Constraints were imposed from the perspective of the responsibility system. The Zhejiang provincial government enhanced the responsibility performance system, combined the river chief system and the intersection water quality assessment, integrated the local government responsibility system and the system of division of responsibility, established the assessment methods for the river chief system, and introduced the river chief system implementation plan for monitoring and evaluation on water environment.

8.3.3 Importance of Systematic and Comprehensive Water Governance by the Government

In the process of water pollution control, Zhejiang actively explored a regional linkage water governance mechanism. For instance, it took coordinated actions in intersectional water environment governance across prefecture-level cities concerning

Tongxiang, Nanxun, Jiande and Lanxi, built a cross-regional linkage cooperation mechanism, i.e. a mechanism for inter-basin coordinated governance of upper and lower reaches, interregional coordinated governance, overseas linkage and interdepartmental linkage, thereby avoiding the "externality" problem to some extent.

Additionally, to solve the three water resource problems concerning "excessive water", "scarce water" and "foul water", Zhejiang not only adopted the "Five Water Co-governance", but also pooled wisdom and strength to give priority to the thumb project of "sewage treatment". Only when the project goes smoothly can the ordinary people give a thumbs-up, and actively devote themselves to water governance so as to facilitate the process of "Five Water Co-governance". Therefore, the systematic governance of public goods carried out by the government is an important part, and even a crucial part determining the success or failure of governance.

China's municipal sewage treatment management lacks a continuous improvement mechanism, the root cause of which lies in a low degree of institutional opening-up. We suggest improving the mechanism and approach for the masses to participate in urban water environment management, and encouraging them to participate in opening-up and public governance. Public governance is an open management mode of public affairs. Adequate social participation can propel the continuous improvement of the municipal sewage treatment management system. Management system opening-up is conducive to building a continuous improvement mechanism, and advancing the continuous improvement of municipal sewage treatment management.

Chapter 9
Case Study on Supply of Environmental Public Goods—Waste Sorting in Shanghai

In June 2018, in a speech at the 18th meeting of the Council of Heads of State of the Shanghai Cooperation Organization (SCO), General Secretary Xi Jinping said, "We should stay committed to the 'Shanghai Spirit'; uphold innovative, coordinated, green, open and inclusive development; pursue common, comprehensive, cooperative and sustainable security; promote open and inclusive cooperation for win–win outcomes; champion equality, mutual learning, dialogue and inclusiveness between civilizations; follow the principle of achieving shared growth through discussion and collaboration in engaging in global governance; surmount difficulties, defuse risks and meet challenges head on." On the forenoon of November 6, 2018, Xi Jinping visited the First Sub-center of Jiaxing Road Sub-district, Hongkou District Citizen Center, Shanghai, and stressed that waste sorting is a new fashion, integrated waste treatment needs public involvement, and Shanghai should make sure garbage management is done well.

9.1 Tough Truth of Household Waste Pollution in Shanghai

Urbanization has been the main driving force of China's economic development since the reform and opening-up, and laid a foundation for further modernization. After entering the twenty-first century, China's development stepped in a new normal. The urbanization process has been no longer limited to the blind pursuit of economic indicators, but gradually turned to new urbanization development with a theme of "Green, Coordination and Sustainable Development". The goal of "speeding up the reform of the system for developing an ecological civilization, and building a beautiful China" presented in the report of the 19th CPC National Congress was the latest requirement for pursuing green and healthy development in China's small, medium and large-sized cities.

J. Yang, *Effective Supply Mechanism and Path of Environmental Public Goods in China*, Public Economy and Urban Governance in China,
https://doi.org/10.1007/978-981-99-2939-9_9

In the process of building green cities and creating a civilized environment, household waste sorting and governance is not only the issue most closely associated with the ordinary people's lives, but also an important and difficult issue of common concern to governments at all levels. In recent years, with the rapid advancement of China's urbanization, the continuous increase of city population and the continuous improvement of the people's living standards, the problem of "waste siege" has become increasingly serious. According to the *Annual Report on the Prevention and Control of Solid Waste Pollution in Large and Medium-sized Cities in 2017* released by the Ministry of Environmental Protection (present-day Ministry of Environment and Ecology), in 2016, 214 large and medium-sized cities of China produced 188.505 million tons of household waste, and disposed 186.844 million tons, with a disposal rate of 99.1%. The harmless disposal rate of China's municipal solid waste has maintained at a high level of above 90%, but the output of municipal solid waste has shown a rising trend year by year, implying that the waste problem has not been fundamentally solved.

In terms of the clean-up amount of municipal solid waste, the annual household waste output in Chinese cities has still maintained a steadily rising trend year by year, with an annual growth rate of 3.9%. From a regional perspective, in 2016, China's top 10 cities in waste output included Beijing, Shanghai, Guangzhou and other megalopolises, the waste output of which accounted for 30% of the total municipal waste output published. This suggests that China's municipal solid waste output is significantly related to the urban development level and the number of urban population. It follows that big cities have increasingly become the hardest hit areas of environmental problems in China, and shows that China has a weakness in waste governance in the green city construction process.

Shanghai is one of representative cities for the most advanced economic level and urbanization development in China. Its history of waste sorting can be traced back to the 1970s. The *Implementation Plan for Classified Collection and Disposal of Municipal Solid Waste in Shanghai* issued in 1999 marked that Shanghai's household waste sorting was officially elevated to government push. Meanwhile, however, the actual progress of Shanghai's waste sorting underwent numerous difficulties, including the difficulty in improving the participation enthusiasm of all sectors of society, and the increase in waste output. In recent years, apart from the short-lived decline of waste output in 2011, the waste output has shown a rising trend year by year. Especially in 2016, the growth rate of Shanghai's household waste output reached 11.4%, which caused the city's household waste output to hit an all-time high of nearly 9 million tons in 2017, ranking second nationwide, next only to Beijing.

According to Fig. 9.1, the municipal solid waste output of Shanghai only had a gentle drop in 2011, but maintained a slightly rising trend in other years. The growth rate of household waste slightly dropped in 2017, but because of the huge base number, its overall development trend remained unoptimistic. As regards this problem, Shanghai has released the *Three-Year (2018–2020) Action Program of Shanghai Municipality for Environmental Protection and Construction*, taken a series of measures to guide and regulate the city's household waste governance process, and continued to draw upon the experience of other cities at home and abroad to

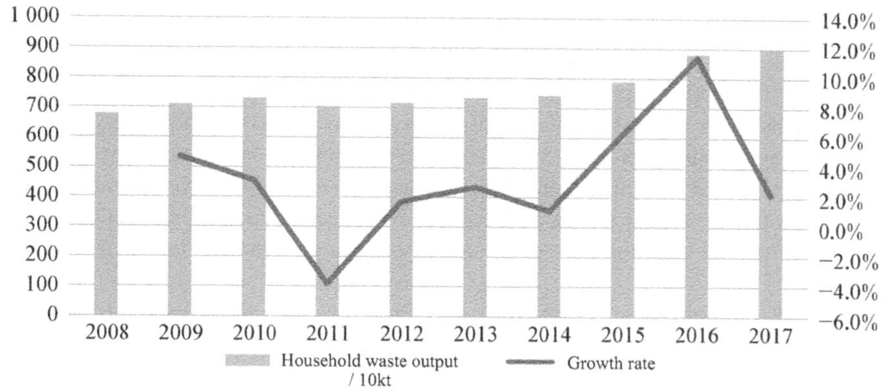

Fig. 9.1 Output and growth rate of municipal solid waste in Shanghai from 2008 to 2017

make attempts, but due to multiple factors concerning small-scale trials, a lack of policy continuity, and residents' weak willingness to cooperate, the effect was fairly limited. The city still has a long way to go to realize effective waste governance.

Therefore, to advance Shanghai's new-type urbanization that is green and sustainable, in 2018, the municipal government issued the *Implementation Plan for Establishing and Improving the Whole-Process Sorting System of Household Waste in Shanghai*, officially put waste sorting into legislation, and tried to use stricter laws and regulations to monitor all aspects of waste disposal by a multi-pronged approach, and to employ various means to encourage all sectors of society to actively participate in waste sorting governance.

9.2 Process of Household Waste Pollution Control in Shanghai

9.2.1 Household Waste Governance Policies of Shanghai

As one of the first cities to propose household waste sorting in China, Shanghai has repeatedly practiced the modes, methods and implementation process of waste sorting, and accumulated certain experience.

Its origin can be traced back to the 1970s when Shanghai implemented household waste sorting in accordance with the standard of compostable and non-compostable waste. This classified disposal mode according to "kitchen waste" and "other waste" became the initial concept of waste sorting in the eyes of citizens.

In the mid-1980s, Shanghai began to advocate the collection of waste in bags; in the same period, Xinhua Sub-district, Changning District, implemented the pilot classified dumping and clean-up of household waste on a small scale, but the effect was insignificant because of a single pilot area.

Before the 1990s, Shanghai's waste sorting was purely dependent on the publicity and appeals from the government and social media, as well as the supervision by neighborhood committees and the citizens' conscious action, without any relevant supporting rules or regulations. In 1995, the Shanghai municipal government took the lead in proposing the harmless, reduced and resource-based disposal of household waste, which becomes the core of Shanghai's waste governance and remains alive so far.

In 1999, the detailed rules for waste sorting in Shanghai were listed into government documents in the form of written text. The details are described in Table 9.1, which lists the detailed rules and specific policies. Afterwards, in the stages of promotion, adjustment and implementation, the detailed rules were greatly adjusted and recompiled.

In the same year, the Shanghai municipal government officially issued the *Implementation Plan for Classified Collection and Disposal of Municipal Solid Waste in Shanghai*, marking that the settlement and execution of Shanghai's waste sorting was officially elevated to the government level. In 2000, Shanghai was listed into China's first batch of pilot cities for waste sorting by the Ministry of Construction (present-day Ministry of Housing and Urban–Rural Development), and promoted

Table 9.1 Waste sorting standards of Shanghai in different stages

Stage	Year	Detailed rules for sorting
Promotion	2000–2003	Residual waste, household food waste, hazardous waste Discarded batteries, waste glass undergoing special sorted collection
	2003–2006	Incineration area: incombustible waste, hazardous waste, combustible waste Other areas: compostable waste, hazardous waste, other waste
Adjustment	2007–2010	Residential areas: hazardous waste, glass, recyclables, other waste Office space: hazardous waste, recyclables, other waste Public places: recyclables, other waste Others: decoration waste, bulky waste, kitchen waste, disposable plastic lunch boxes, etc. undergoing special collection, transportation and disposal
	2010–2011	Separation for bulky waste: decoration waste, unit kitchen waste, bulky waste, dry branches and fallen leaves, etc. Classification for small ones: hazardous waste, glass, waste clothes, food waste, other waste, etc
	2012–2013	Recyclables, hazardous waste, kitchen waste (household food waste), other waste (residual waste)
Implementation	2014 to the present day	Residual waste, household food waste, hazardous waste, recyclables

waste sorting in 600 housing estates of the central urban area. Since that year, the municipal government has launched the "Three-Year Action Programs for Environmental Protection and Construction". See Table 9.2 for specific policies. In the seven rounds of action programs, the government defined the goals for each round and introduced supporting policies. It could be observed that the Shanghai municipal government listed "strengthening solid waste disposal" as one of the important governance policies in each round of action program. Focusing on the outline of a series of three-year action programs, the municipal government developed specific implementation rules, measures and plans for household waste, including the *Measures of Shanghai Municipality for Promoting Household Waste Sorting and Reduction* in 2014, the *Implementation Plan of Shanghai Municipality for Mandatory Sorting of Unit Household Waste* in 2017, and the *Implementation Plan for Establishing and Improving the Whole-Process Sorting System of Household Waste in Shanghai* in 2018, hoping to propel continued public attention to this problem and ensure the smooth progress of household waste sorting.

Table 9.2 suggests that in the first round of "Three-Year Action Program for Environmental Protection and Construction", Shanghai has listed "solid waste disposal" as one of the five major areas to be propelled. With the continuous improvement of municipal solid waste governance, governance goals continued to be refined. On the basis of consolidating the harmless disposal level of garbage, Shanghai introduced the thought of circular economy, increased the resource utilization percentage of household waste, and raised higher requirements for the construction of disposal facilities and capacity. The *Action Program of Shanghai Municipality for Building the Whole-Process Sorting System of Household Waste (2018–2020)* issued in April 2008 uses two chapters to introduce a detailed plan for a new round of goals and requirements for household waste governance, and specifically lists a number of construction projects for harmlessness and resource utilization of household waste. The increase in writing length indicated that the municipal government attached unprecedented importance to the governance of household waste. Previously, the *Implementation Plan for Establishing and Improving the Whole-Process Sorting System of Household Waste in Shanghai* issued in March 2018 raised the requirement for fully implementing the mandatory sorting of household waste, ensuring the orderly progress of waste sorting by institutional and administrative means.

9.2.2 Pilot Sorting of Household Waste in Shanghai

Since 2007, Shanghai has carried out the citywide pilot promotion and demonstration of waste sorting in housing estates. Through over a decade of efforts, in 2018, the pilot scope was extended from the initial 100 housing estates to more than 10,000 housing estates. According to survey findings, the participation of residents in household waste sorting significantly increased in pilot housing estates, and residents' sorting awareness was significantly enhanced, but some problems remained,

Table 9.2 Waste governance policies of "Three-Year Action Programs for Environmental Protection and Construction" of Shanghai

Stage	Round	Goal	Specific policy
Promotion	1st round: 2000—2002	Focus on advancing solid waste disposal	*Implementation Plan for Classified Collection and Disposal of Municipal Solid Waste in Shanghai*
	2nd round: 2003—2005	Speed up the classified collection of household waste, with the classified collection rate of municipal solid waste up to 60%	
	3rd round: 2006—2008	Improve the household waste collection and transfer system, and build a batch of integrated disposal facilities, with the harmless disposal rate of municipal solid waste up to 80%	
Adjustment	4th round: 2009—2011	Accelerate the construction of facilities for collection, transfer and disposal of household waste, optimize the treatment and disposal structure, and realize the containerized airtight transport, with the harmless disposal rate of municipal solid waste up to 85%	
	5th round: 2012—2014	Fully advance the classified collection of household waste, improve corresponding classified collection, transport and disposal systems, and propel the safe disposal and up-to-standard discharge of landfill leachate, with the harmless disposal rate of municipal solid waste up to 95%	*Measures of Shanghai Municipality for Promoting Household Waste Sorting and Reduction*
Implementation	6th round: 2015—2017	Speed up the process of harmless disposal and reduction of household waste. Continue to improve the "whole-process sorting system", expand the implementation areas of waste sorting, and continue to improve the resource utilization efficiency and standardized management level of waste sorting. Enable the sorting and reduction of municipal solid waste to cover 5 million households, maintain the terminal household waste disposal at 0.62 kg/day for each person, and increase the disposal capacity of household waste by more than 5000 t/d	*Implementation Plan of Shanghai Municipality for Mandatory Sorting of Unit Household Waste*

(continued)

Table 9.2 (continued)

Stage	Round	Goal	Specific policy
Implementation	7th round: 2018—2020	In the context of focusing on solving the current insufficient disposal capacity and other outstanding problems, consolidating harmless results, breaking the bottleneck of reduction, opening up resource utilization channels, and basically completing a systematic classified collection, transport, disposal and recycling system for solid waste, continue to propel the sorting and reduction of household waste, and improve the terminal disposal system of household waste. By 2020, strive to enable the recycling rate of household waste to reach 38%; the harmless disposal rate of household waste to reach 3.28×10^4 t/d; the classified disposal capacity of household food waste to reach 7,000 t/d	*Implementation Plan for Establishing and Improving the Whole-Process Sorting System of Household Waste in Shanghai*

including different implementation effects in pilot areas, insufficient hardware facilities for municipal solid waste, and a large gap in the participation willingness of residents of different ages.

The latest *Survey Report on Household Waste Sorting Status of Pilot Housing Estates in Shanghai* suggests that through the continuous media publicity and a series of incentive measures taken recently, in the 51 pilot housing estates of 16 districts, 90% of residents have known the "Four-Category" separation requirements of household waste, and supported waste dumping at a fixed time and place; more than half of the residents participated in the Green Account campaign; meanwhile, nearly 30% of pilot housing estate residents developed the habit of active sorting. This shows that through continuous policies and media publicity, most residents have heard about the detailed rules and standards for household waste sorting, and gradually attempted to put them into action.

However, in the context of the increasing popularity of waste sorting, these pilot areas greatly varied in implementation degree. As regards the participation in the Green Account, the participation rate of residents in Hongkou District was 81.8% while that in Yangpu District was only 27.7%, with a gap up to 54.1%; in terms of the standard setting of dustbins, Changning District reached 48.7% while Jiading District only realized 5.4%, with a gap of 43.3%; as for the sorting accuracy, the accuracy rate of Chongming District was 61.0% while that of Xuhui District was 45.9%, with a gap of 15.1%.

According to survey findings, the construction of waste sorting hardware facilities in the city still has plenty of room for improvement. For the setting of dustbins, the four-category dustbins only accounted for 21% of the total number of bins, and the two-category waste bins accounted for the highest proportion, 36%. Besides, there is a certain gap in the completeness of waste disposal facilities between pilot and non-pilot housing estates. 25.4% of pilot housing estates have used the four-category dustbins for collection while only 2.0% of non-pilot housing estates have used such bins.

Additionally, the survey suggests that residents aged 61–69 are the main force of household waste sorting. The young people aged 18–25 are weak in the awareness of waste sorting, and the "shirking of trouble" is the principal factor influencing their action; besides, these people have a lack of leisure time for "indoor sorting" because of busy work and study, which is also an important factor influencing their enthusiasm for participation. The survey also indicates that most residents believed that giving full play to the role of neighborhood committees and property management is the key to implementing waste sorting. Thus, how to achieve the orderly progress of waste sorting "from top to bottom" and "from bottom to top" by the combination of policy making and primary-level governance is the core of the current work.

9.2.3 Core Mechanism of Household Waste Governance in Shanghai

With the profound changes in the social organization structure, social ties and state-society relations, the transformation from "strengthening social management" to "innovating on social governance" has begun to be realized at the state level, and the new requirement for innovating on social governance has been put forward. The report of the 19th CPC National Congress proposes establishing a social governance model based on "collaboration, participation, and common interests" so as to emphasize the core concept of multi-subject participation in social governance. Moreover, the CPC Central Committee further required shifting the focus of social governance to the community level, emphasized the role of social organizations in urban and rural communities. Obviously, primary-level organizations are playing an increasingly significant role in public affairs.

1. Subject of waste governance—residents' self-governance + government supervision and incentives

Household waste governance is an issue integrating the attributes of both private and public interests. The current situation suggests that waste governance is not only advanced by the government through policy making, but also dependent on residents' conscious cooperation. As residents are the actors of waste discharge and sorting, their behavior can directly influence the effect and level of governance. Meanwhile, as a life-oriented problem, it has certain chanciness. Residents' behavior is highly susceptible to policy incentives and other external factors. As a saying that goes, "Collective action is not a natural phenomenon, but a social construction, which is a chance created, invented and determined by actors by utilizing their unique resources and capabilities." The role of the government and social organizations (enterprises) cannot be underestimated either. On this basis, this chapter selects the direct actors of waste sorting—residents—as the main object of study, brings the external effect of the government and social organizations (enterprises) into the system, and proposes the following two viewpoints.

Viewpoint 1: Establishing a primary-level system of "residents' self-governance" is an important foundation for propelling waste governance.

The key to the success of waste governance is residents' active action. As rational persons, residents tend to weigh the costs and benefits of waste sorting. However, waste sorting is a complicated task. Residents' indoor sorting will directly cause the rise of their time costs, and make it difficult to obtain actual returns. Thus, how to lower the cost of waste sorting and improve residents' sense of gain becomes the core for effectively advancing waste governance. In light of the importance attached to the improvement of grassroots organization in the report of 19th CPC National Congress, this chapter further proposes the waste governance mode of "residents' self-governance". The relational structure of the acquaintance society, as well as non-institutionalized elements concerning human relations and face saving could be utilized to conduct self-supervision and self-motivation on waste sorting; through the interaction and daily influence between neighbors, "passivity" could be transformed into "initiative", and community competition and appraisal could be employed to improve residents' sense of gain.

Viewpoint 2: The government and social organizations (enterprises) play a role in supervising and stimulating waste governance.

The "residents' self-governance" system mainly in reliance upon interpersonal interaction has certain vulnerability and uncertainty, thus "rigid constraints" need to be imposed by institutionalized means so as to set a basic role model for residents and to supervise them. The government may strictly regulate waste governance by introducing laws and regulations, and innovate on reward and punishment measures, e.g. learning from the German mode, adopting a "collective" punishment mechanism, raising the waste clean-up fee of an entire area in case any individual living in that area fails to cooperate in waste sorting; or learning from the Japanese mode, building waste disposal facilities in communities to internalize external environmental problems, and to turn residents into direct stakeholders. Additionally, social organizations and enterprises may take advantage of their professionalism and funds to give residents material incentives, and cooperate with the government in joint governance.

2. Residents' motivation for participating in waste governance

The motivation theory means a theory with regard to the generation and mechanism of a motivation, as well as its relations with needs, behaviors and objectives. The concept of motivation is derived from psychology, and mainly manifested as a subjective desire or intention to pursue an objective, which is a kind of human self-consciousness for pursuing an intended purpose. When people's needs reach some intensity and there exists an object meeting such needs, needs can be translated into motivations. Based on this theory, with residents' motivation for performing waste sorting obligations at the core, this chapter divides motivations into endogenous and exogenous motivations, and researches the effect thereof on residents' behavior and the interaction between them.

(1) Endogenous motivation

Residents' endogenous motivation for waste sorting, i.e. residents' conscious participation in waste governance, stems largely from their desire to improve the living environment and consideration of their own social benefits. In the past practice of waste governance, residents lacked enthusiasm for participating in community-level public affairs resulting from the long absence of social citizenship consciousness in China. Due to their indifference to community affairs, their individual rationality prevailed. As a result, when facing choices, they tended to give priority to maximizing their own interests, but neglected the gains or losses of group interests. Moreover, community governance is non-exclusive, so the majority of residents are often willing to be "free-riders", and unwilling to make contributions to community affairs, with a wish to directly enjoy the fruits of governance. This spectator thought finally will lead to unclear responsibilities for waste sorting and less rigorous execution, and overburden sanitation authorities.

Therefore, to effectively improve residents' enthusiasm for participating in waste sorting, a "residents' self-governance" system should be established so as to enable improving the overall regional environmental level to be the common goal of community residents. Based on the acquaintance society, "residents' self-governance" relies on close interpersonal relations, but not strict rules or regulations; on the basis of traditional non-institutionalized elements, modern "residents' self-governance" incorporates the thought of primary-level governance, and introduces primary-level management organizations concerning residential buildings and neighborhood committees to assist in governance. This management system combining institutionalization and non-institutionalization enables residents to be fully influenced by families, neighborhood and social environment, and boosts their sense of belonging in and acceptance of communities, thereby, enhancing the connection between community interests and residents' individual interests.

(2) Exogenous motivation

The exogenous motivation for waste governance mainly derives from the government's ex-ante policy making, interim regulation, and ex-post reward and punishment measures, and includes the assistance and coordination role of social organizations (enterprises). The ultimate goal of exploring exogenous motivations is to stimulate and consolidate endogenous motivations. At present, the government is dominant in the public governance of China, thus the effective implementation of its duties and powers is an important premise of policy advancement. The formation of residents' endogenous motivation for waste governance needs to be guided by a clear goal, and the government is a powerful maker of this goal. The government employs supervision and reward and punishment means, draws support from corporate funds and human resources of social organizations, and takes incentive measures to influence and change residents' behavior, thereby enabling them to consciously participate in waste sorting on the basis of a clear direction of efforts. Furthermore, the punishment means adopted by the government and related departments according to laws and regulations can increase the violation cost to some extent. On the premise that the violation cost is higher than the benefit from environmental improvement, as rational persons, residents can actively cooperate with relevant parties in performing waste sorting after weighing advantages and disadvantages.

(3) Interactive relations between endogenous and exogenous motivations

The endogenous motivation for waste governance, namely the strong willingness to improve the living environment is the root reason for changing residents' behavior, but still needs cooperation with exogenous motivation for stimulation. The government achieves a close combination of residents' private interests and public interests by establishing rules and regulations, enables residents to further take group interests into account on the basis of individual rationality, and spurs them to enhance their subjective impetus for participating in waste sorting. The interaction between endogenous and exogenous motivations improves the level of waste governance step by step.

Additionally, endogenous and exogenous motivations constitute a loop of "execution–feedback–adjustment–re-execution–second feedback", as shown in Fig. 9.2. Based on the feedback principle, in the existing waste governance practice, the government first developed the goals and norms for waste sorting, and worked with relevant units and enterprises to tease out the process and key parts of waste governance. Residents executed waste sorting in accordance with the goals and requirements, and relevant units gave feedback after making management statistics to help the government timely adjust the goals in the next stage according to the status and problems identified. Meanwhile, residents might give feedback on the incentive behavior of relevant enterprises to enable enterprises to make timely adjustments according to residents' willingness. In the continuous loop of "feedback" and "execution", residents would gradually foster their awareness of waste sorting, namely, an endogenous motivation.

9.3 Typical Case: "Government Guidance, Residents' Self-governance, and Multi-party Cooperation" Mode of Changning District, Shanghai

As a national civilized district, Changning District has ranked among the top of the city in terms of waste sorting. From taking the lead in the pilot implementation of the "Green Account" scheme in 2014, to implementing the "Dual-Network Cooperation and Integration" with respect to classified collection and transportation of household waste and renewable resource recycling, and the waste dumping at a fixed time and place in a sorted manner in 2017, then to becoming one of the city's first batch of pilot districts for household waste sorting in 2018, Changning has provided valuable experience for Shanghai to vigorously propel the "reduction, resource utilization and

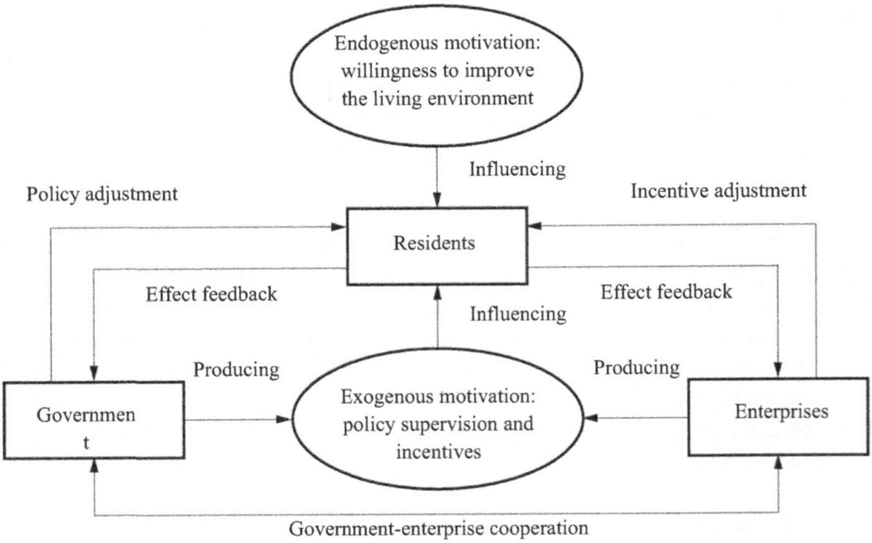

Fig. 9.2 Interactive relations between endogenous and exogenous motivations

harmlessness" of household waste. In the process of implementing waste sorting and reduction, with the aim of insisting on the "guidance of Party building and multi-party cooperation", under the concerted efforts of the government, enterprises and residents, Changning improved all social parties' enthusiasm for waste disposal, and made gratifying achievements by combining policy guidance and incentive measures.

9.3.1 Reasons for Changning District Selected into the First Batch of Pilot Districts for Household Waste Sorting

The Shanghai municipal government selected Changning into the city's first batch of pilot districts for household waste sorting for the following three reasons.

First, Changning has been selected as a national civilized district more than once, and the overall quality of residents therein is high, which is conducive to the promotion and implementation of the pilot work. Changning was awarded the title of "National Civilized District" in 2011, and continued to retain the title after successfully passing the reexamination in 2017. The district had 10 sub-districts (towns) rated as municipal civilized communities, and successfully built a number of city-level or district-level civilized housing estates and units. In addition, it was reconfirmed as a national sanitary city (district) in 2017. To smoothly pass multiple rounds of appraisals and reviews by the central inspection group, all sub-districts and units in Changning conducted vigorous publicity and took active action, not only publicized civilized district construction by multiple channels and mobilized units and residents to improve the city environment, but also strengthened supervision and inspection, and strictly punished those units and individuals for uncivilized behavior. As one of the important indicators for selecting national civilized districts and national sanitary cities (districts), household waste governance has been a focus of social concern and a core task of the Changning district government. The effective advancement of waste sorting could further consolidate and improve the image of the district, and play an exemplary role in facilitating the improvement of civilized level and city environment of other districts.

Second, Changning carried out pilot waste sorting in some communities in the 1980s, and thus had a certain popular base. In 1984, the district government implemented the pilot classified dumping and clean-up of household waste in Xinhua Sub-district; in the second year, the pilot work was popularized on a small scale in a sub-district designated by respective districts of the city. In that period, the waste sorting principle basically followed the classification method of "being compostable or not" advocated in the 1970s. Residents dumped vegetable and fruit peel, nut shells, coal dust, etc. into green dustbins; scrap iron, glass and the like into reddish-orange bins; and construction waste in a designated centralized waste collection site. After being sorted by relevant units, compostable organic household waste was directly transported by sanitation departments to fields as compost; scrap iron, waste glass, and other recyclable waste were directly transported to reclamation depots; bulky construction waste was cleaned up upon further sorting. The effect was less than ideal because of the small scale, but the pilot work imperceptibly fostered residents' awareness of waste sorting to some extent, and laid a solid foundation for the implementation of present-day larger-scale, more standardized and stricter household waste sorting.

Third, with a complicated population structure, numerous enterprises and public institutions, as well as densely-distributed key functional areas, Changning is a representative district in the city. As of the end of 2017, Changning had a permanent population of 693,700, of which non-native permanent population accounted for more than a quarter; expatriates accounted for 10.29%, close to one fifth of the total number of expatriates in the city. Compared to the publicity and guidance of household waste sorting targeting the non-native population, the community work targeting expatriates appeared more difficult. To solve language barriers, cultural gaps, differences in customs and habits, and so forth, Changning required relevant personnel to master certain foreign language skills, and have some understanding of the cultures of other countries. As an international metropolis inhabited by locals, non-locals and foreigners, Shanghai ought to enhance the participation of non-locals and expatriates in the process of implementing household waste sorting. If this can achieve certain results during practice in Changning, it will accumulate precious experience for elevating the internationalization level of Shanghai's community work. Besides, In Changning, there are a number of major functional areas, including Hongqiao Business District, Hongqiao Development Zone, Airport Economic Park, and Zhongshan Park Business District; newly-built housing estates and old alleys coexist; the secondary and tertiary industries are distributed crisscross. The complicated composition is a huge challenge to the implementation of household waste disposal. If a plan could be developed and advanced according to the characteristics of each area and local conditions, then reference could be provided for the work in other areas of the city, and a breakthrough in waste sorting would be finally made in all areas.

9.3.2 Implementation Mode of Household Waste Sorting in Changning District, Shanghai

In accordance with the aim of the *Implementation Plan for Establishing and Improving the Whole-Process Sorting System of Household Waste in Shanghai*, Changning established a waste governance mode with "Government Leading, Resident Cooperation, and Enterprise Assistance" at the core, fully motivated enterprises, public institutions and residents under the leadership of the government and the concerted efforts of nonprofit social organizations, and employed policy means to strictly supervise and manage the process, thereby, achieving the continuous and effective progress of household waste sorting.

1. Government leading and institutional guarantee

To propel waste sorting and reduction, first, the government should play a role in leading by example and policy guidance. According to the requirements of the *Action Program of Shanghai Municipality for Building the Whole-Process Sorting System of Household Waste (2018–2020)* issued by the municipal government in combination with its own characteristics, Changning started with hardware and software to steadily advance waste governance by a double-pronged approach.

To have a full understanding of the practical problems in waste governance of housing estates, and to put forward solutions according to local conditions, Changning District proposed that importance must be attached to the combination of commonality and individuality in waste sorting; the characteristics of different housing estates should be considered in the reproduction and promotion process of successful experience modes so as to avoid "sameness". Thus, the sub-districts in Changning convened survey symposiums with the presence of Party organizations in residential areas, owners' committees, property management companies and residents' representatives more than once, determined "a plan for each housing estate respectively", listed the major commonality and individuality problems to be solved by communities in the front part of waste sorting according to such basic information as the number of households in each housing estate, the composition of residents, the dustbin rooms, and the foundation of public opinion, and developed corresponding detailed plans with regard to waste dumping time, waste transport time, dustbin room design, as well as the number of volunteers on the basis of the overall plan determined by the district government.

In terms of improving hardware facilities, the district government required all sub-districts to endeavor to improve the surrounding environment of dustbin rooms, and increase residents' satisfaction with waste collection and treatment facilities. For instance, according to the status surveyed, Chengjiaqiao Sub-district presented the dustbin room "beautifying, addition and cleaning" project: "beautifying" appearance; "adding" fountains, canopies, lounges and other facilities; ensuring a "clean" environment. In addition, the style of dustbin rooms was consistent with that of housing estates to ensure "each dustbin room with unique characteristics"; "rest stops" might be arranged beside dustbin rooms if conditions permitted to provide services for road cleaning personnel in rest time.

In terms of innovating on the software system, the district government implemented and advanced the "Dual-Network Cooperation" of waste sorting and recycling, and explored the government-led "Five Fixed" ("Fixed Time, Fixed Place, Fixed Category, Fixed Personnel, and Fixed Systems") work mode for household waste. The specific content included: arranging each staff member with a dustbin room, advocating the principle of localized staff placement; building a team of volunteers composed of veteran officials, Party members, residents' representatives, et al. in the district to assist residents in accurate classification; developing systems concerning regular meetings, training, ledger registration, and data analysis to standardize the process of waste sorting.

To ensure the effective implementation of waste governance plans, all sub-districts analyzed the waste reduction line on a regular basis to master the plan progress in real time, and conducted assessments. In the survey process, sub-districts carried out daily statistics, weekly analyses and monthly summaries on the waste sorting of each pilot housing estate, summarized the fluctuation trend of the waste reduction line for each housing estate in every stage, and analyzed the reasons thereof. Through summarizing and analyzing data, sub-districts focused more on the rebounding stage of the waste reduction line in a targeted way in the subsequent pilot work, and performed re-mobilization, re-launching and re-consolidation.

2. Resident cooperation and government leading

To fully arouse residents' initiative in and enthusiasm for waste sorting, Changning gave play to the exemplary role of Party members, held a variety of activities through sub-districts and neighborhood committees, and imperceptibly fostered residents' environmental awareness by means of education through lively activities. For example, in Hongxu Housing Estate, Xianxia Sub-district, residents used waste materials to set up a "Bottle Vegetable Garden"; held the "Appraisal for the Best Family in Waste Sorting"; as well as such characteristic activities as the Lantern Festival Riddle Games on Waste Sorting Knowledge, the Waste Sorting Training with Cooking Display, and the Gifts for Scanning QR Code of Waste Sorting. For another example, during the practice, Hongqiao Sub-district mobilized Party members and officials of government agencies, as well as Party members working at enterprises and public institutions in the district to take the lead in participating in housing estate waste sorting, organized backbone personnel to take action first, and explained "three questions": Why should we implement waste sorting? Why should we throw away garbage at a fixed time and place? Why should we advocate "Dual-Network Cooperation"? Furthermore, the sub-district introduced social organizations, invited "Green Housewife" and other organizations to offer an environmental protection lecture tour and to provide support for detailed guidance on waste sorting, and gave residents professional guidance on waste sorting through Ten Lessons on "Waste Sorting", "One Square Meter" of Vegetable Garden, "Family Subtraction Life", "Challenging the Environmental Protection Experts", and other lively and interesting projects.

According to the characteristics of each housing estate, sub-districts worked out a set of waste sorting operation manuals for different types of housing estates to guide residents in waste sorting. For instance, as regards Deluxe Family Housing Estate where many foreigners lived, Ronghua Neighborhood Committee distributed multilingual publicity packages at their doors, and conducted training on domestic workers; for housing estates without a dustbin room, Hongqiao Sub-district developed a detailed plan for pilot fixed-time collection and transport of classified waste in housing estates, provided transportation support in conformity with the requirements for classified clean-up, and strengthened clean-up on the basis of waste sorting and collection by property management.

3. Enterprise assistance, filling vacancies

In addition to strengthening household waste governance in housing estates, the enterprises and public institutions throughout the district are a key part of governance as well. Playing the dual roles of "intermediaries" and "participants", they could assist the government in jointly advancing the effective implementation of waste governance.

As "intermediaries", enterprises could effectively eliminate the gap between policy making and actual implementation, and employ market-oriented means to fix loopholes and deficiencies in government policies. For example, to solve the "difficulty in redemption" of green account points, Huizhong Green Fortune Development and Promotion Center cooperated with Carrefour and other businesses to jointly establish the "Green Alliance" project based on the green account, held the Green Account Carnival, and worked with Alipay on "Double Twelve" sales promotion, thereby enabling the green account to have continuous exchange rights and interests guarantee. To solve the insufficient waste governance capacity, the district government actively introduced private enterprises, e.g. Shanghai Xinjinhua Commercial Co., Ltd., to undertake the cleaning and renewable resource recycling of bin rooms in the district, thereby enabling social professionals to assist in elevating and improving waste governance capacity.

As "participants", corporate parks have strengthened the renovation and publicity of waste disposal facilities, allowing the concept of waste sorting to take firmer root among the masses. For instance, in the Airport Economic Park, the first pilot unit for mandatory sorting of household waste, not only the dustbins in over 70 pantries of the office park have underwent a renovation pursuant to the sorting standard of residual waste and household food waste, but also white-collar workers have changed their garbage throwing habit from direct throwing to throwing after thinking it over. In addition, the park has established a seamless waste collection and transport system: floor cleaners check the residual waste and household food waste bins of each department; park cleaners classify the recyclable waste in residual waste bins, and record the weight of each bin of waste into an account book, waiting for the residual waste and household food waste trucks from sanitation departments for separate transport.

9.3.3 *Implementation Effects and Existing Problems of Household Waste Sorting in Changning District, Shanghai*

At present, Changning has achieved certain results in hardware facility construction and daily supervision of household waste treatment, but still has room for improvement.

1. Significant effect of waste sorting

As of the end of 2018, Changning had a total of 726 housing estates, of which 581 housing estates had a total of 907 household waste bin rooms, but 145 housing estates had not been equipped with waste disposal facilities; half of these housing estates had established "Dual-Network Integration" service sites, and achieved the integrated operation of the household waste sorting, collection and transport system and the renewable resource recycling system. As the foundation of household waste sorting, the classification of residual waste and household food waste in all housing estates of Changning has produced initial results. From May to December 2018, the output of household food waste increased from 187 t/d to 314 t/d while that of residual waste decreased from 814 t/d to 629 t/d, which showed that with the orderly advancement of the publicity and guidance of household waste sorting, residents increasingly improved their waste sorting accuracy. Besides, the resource utilization rate of household waste was significantly improved as well, from 27.45% in May 2018 to 43.7% in December 2018, which reflected a breakthrough in waste recycling.

Meanwhile, to deepen the practice effect of waste sorting governance, Changning developed implementation plans for different types of housing estates, and carried out work according to local conditions. Zhoujiaqiao, Hongqiao and Chengjiaqiao became the first batch of sub-districts to apply for building demonstration sub-districts of household waste sorting. The waste disposal and sorting data of Chengjiaqiao Sub-district and all housing estates therein from June to November 2018 was abstracted to analyze its effects and deficiencies in the advancement process.

Within half a year, regarding the housing estates participating in "dumping at a fixed time and place", the advancement scale extended from the initial six pilot housing estates to an "entire area" comprising 27 housing estates under eight neighborhood committees. With the continuous expansion of the participation scale and the continuous increase in the number of covered residents, the governance mode continued to be optimized and mature. In the context of a substantial increase of pilot housing estates participating in waste sorting, the resource utilization rate of household waste declined gently from 51% in June 2018 to 45% in November 2018; the accuracy rate of classified disposal maintained above 90%; the "bag-free" disposal rate of household waste also maintained at a high level of above 90%, and Shanghang New Village as the first to participate in the pilot work within the sub-district even reached a record high of 100%. This not only showed that through half a year of habit formation, community residents have kept waste sorting knowledge in mind, and could actively cooperate with community staff to participate in relevant work; besides, through the experience accumulation and summarization in the early stage, those housing estates newly joining the pilot work could implement the waste sorting system in a more efficient fashion.

Comparing the data of October and November 2018, we could further find that though three housing estates newly joined the pilot sorting work, the total waste output in the pilot area reduced by 18,201.22 kg; the proportion of total output of residual waste declined by 2.2%, the proportion of that of household food waste rose by 2.97%, the proportion of recyclables rose by 0.77%, and the increment of recyclables and household food waste was 1.54 higher than the decrement of residual waste, which suggested that the waste reduction effect of Chengjiaqiao gradually improved. Additionally, the resource utilization rate of waste had a slight increase of 2%, which showed that guided and driven by prior pilot housing estates, the residents of newly-joined housing estates and relevant responsible persons were capable of familiarizing themselves with the detailed rules for waste sorting more rapidly, and performing secondary sorting for recyclable waste.[1]

2. The green account popularity of waste sorting to be improved

The green account is a waste sorting point scheme implemented by the Shanghai municipal government, with an aim to encourage residents to actively participate in waste sorting by means of exchanging points for daily supplies. As of November 2017, the overall green account card swiping rate of Chengjiaqiao Sub-district was 71.7%. Through half a year of efforts, the prior pilot housing estates in the sub-district, e.g. Shanghang New Village and Jiali Haoyuan, have been capable of maintaining the monthly card swiping rate at a high level of above 90%, and sometimes might even realize a rate of 100%. This shows that residents have developed a habit of household waste sorting. For those newly-joined housing estates, e.g. Baobei Neighborhood Committee, their residents' willingness to swipe the card was relatively low. This shows that the development of a waste sorting habit is a gradual process. Some housing estates have not been equipped with a card reader, which shows that in the process of propelling municipal solid waste governance, uneven regional implementation still exists, and the experience of prior pilot housing estates needs to be used for reference so as to improve the overall waste sorting level.

Chengjiaqiao is one among the first batch of demonstration sub-districts, where 42 housing estates have achieved the classified dumping of household waste at a fixed time and place, with a resident support rate of above 95%, an accurate disposal rate of 98%, a "bag-free" disposal rate of 95%, and a daily average resource utilization rate of 40–60%. With the advancement of household waste governance, more and more residents have realized the importance thereof, and actively participated in

[1] Data comes from the integrated NGO data.

daily self-supervision to achieve the effective combination of "top-to-bottom" and "bottom-to-top" advancement of waste sorting.

3. Improvement orientation

With the efforts of residents and staff members in pilot housing estates, Chengjiaqiao Sub-district is advancing household waste sorting in an orderly fashion, and has achieved encouraging results. Even so, in the advancement process, there is still room for further improvement. For example, the resource recycling rate remains low; the sorting habit of newly-joined housing estates needs to be developed; the green account popularity needs to be improved. The settlement of these problems needs not only government guidance, but also the participation of enterprises in providing various personalized services with respect to waste governance so as to optimize the overall governance process, and serving as a bridge between the government and residents so as to promote the effective settlement of waste sorting problems.

9.4 Status and Prospect of Household Waste Sorting in Shanghai

Through summarizing the variation trend of the city's household waste output in recent years, the adjustment and implementation of municipal solid waste governance policies over the years, as well as the results of resident questionnaires and surveys, we could find that though the government has advanced waste sorting governance through the "*Three-Year Action Programs of Shanghai Municipality for Environmental Protection and Construction*", and achieved certain results in enhancing residents' environmental awareness, and improving the participation in household waste sorting, but still has many problems to be solved in the overall measure implementation process. These problems are mainly focused on "total quantity control", "policy making" and "social participation".

9.4.1 Problems of Household Waste Sorting in Shanghai

1. Increase of total waste output

The statistics on the clean-up amount of municipal solid waste over the years suggested that Shanghai's waste reduction was unsatisfactory, and the total waste output still showed an increasing trend year by year. In recent years, the municipal government has put in lots of human, material and financial resources, and has completed and is building a number of larger-scale environment-friendly disposal facilities, e.g. the Shanghai Laogang Waste Incineration Power Plant (Shanghai Laogang Landfill Gas-to-Energy Plant) Project, and the Tianma MSW Plant (Tianma MSW Comprehensive Utilization Center) Project, on the basis of consolidating and improving the existing household waste disposal facilities, but the construction speed and scale still could hardly meet the demand for disposing the rapidly increasing household waste output in the city.

As regards the reasons, on the one hand, the continuous improvement of citizens' living standards has caused a rise in the quantity of consumer goods packaging materials, shipping packaging materials, kitchen waste and other waste, and the idle items resulting from excessive consumption have led to an increase of waste output, which could be hardly mitigated; on the other hand, as the development of the city's circular economy is in the preliminary stage for the time being, the resource recycling industry is underdeveloped, and lots of household waste and idle items fail to be recycled in a full and effective way. The city size is positively correlated with the household waste output, but purely improving the disposal capacity and volume of waste disposal facilities to deal with the constant flow of household waste is not an effective solution to the city's "waste siege" plight. To enable the urban ecological environment to embark on the path of sustainable development, the government should develop policies in connection with circular economy and resource utilization industries, guide enterprises to actively devote themselves to related industries, start with "reduction" and "disposal", and combine unclogging and blocking means to strengthen municipal solid waste governance.

2. A big gap in waste disposal capacity

At present, the gap in Shanghai's household waste treatment capacity is mainly manifested in insufficient overall disposal capacity and backward technology.

On the one hand, Shanghai has not satisfied the predetermined requirements for the construction of municipal solid waste disposal capacity. According to the requirements of related action programs, Shanghai should fully realize the harmless disposal of household waste, with the disposal capacity of 3.4 × 104/d. However, as of 2017, Shanghai's daily disposal capacity of household waste was only 2.4 × 104 t/d. The unsatisfactory waste disposal capacity caused the occasional occurrence of illegal waste dumping, the illegal disposal of household waste and other prominent problems. Moreover, the negative externality resulting from such problems even severely damaged the ecological environment of surrounding provinces and cities, and produced a serious negative impact on the overall coordinated development of the ecological environment of the Yangtze River Delta.

On the other hand, Shanghai's disposal capacity of household food waste and residual waste is in a "tight balance" state. The existing harmless disposal technology is difficult to eliminate the impact on the ecological environment. So far, the harmless disposal rate of household waste has been close to 100%, but affected by technology, funds, talents and other factors, the actual disposal level is not high. The disposal modes including landfill and incineration currently in extensive use still have an impact on the ecological environment to some extent, and the most prominent problem is leachate pollution. For example, Anting Household Waste Disposal Plant has excessively discharged leachate into the sewer network for long; the closure of Gu Village Dumping Site in Baoshan District and Zhujiajing Dumping Site in Minhang District has not been completed after over 10 years of shutdown, causing leachate to flow into off-site storm sewers or to be directly discharged into rivers. The deficiency in disposal technology could not only affect the city appearance and residents' feelings, but also easily bring about ecological environment crises and hinder the harmonious development of the city and the nature.

3. Weaknesses in classified transportation of waste

Though in recent years, Shanghai has actively publicized and encouraged citizens to take an active part in waste sorting, the contradiction between classified disposal and classified transportation has existed for long in the actual transportation and disposal process. The causes of this contradiction lie in unclear rights and responsibilities and government fund undersupply in waste governance.

On the one hand, the government's deficiency in defining residents' responsibilities for sorting and relevant sanitation departments' lack of regulation enforcement capability have caused municipal solid waste sorting to be in a disordered state. In terms of Shanghai's waste sorting governance in past years, household waste sorting was mainly dependent on residents' self-consciousness, and had no strict mandatory regulations; moreover, in the daily waste clean-up process, sanitation departments could only verbally notify and educate those residents or units refusing to perform waste sorting, and had difficulty in taking effective coercive measures against any residents or units refusing to perform waste sorting or failing to properly classify waste. The indistinct definition of relevant responsibilities caused sorting quality to be difficult to reach the standard for classified collection and transportation of waste, which not only increased the secondary sorting workload of clean-up units, but also affected the efficiency of waste clean-up.

On the other hand, the rapid increase in the number of pilot housing estates has brought pressure regarding classified transport vehicles and governance funds. This problem is mainly related to the government's working quota calculation in the field of service procurement. At present, according to the requirements for daily waste removal raised by sanitation units, the government collects and transports household waste twice a day, and allocates funds in accordance with the classified transportation standard of residual and household food waste bins, rather than the actual content and quality of waste sorting. The latter obviously needs to rely on clearer definitions of rights and responsibilities, stronger regulatory capability and residents' self-consciousness, which can be counted as a huge challenge for urban sanitation work. Under the circumstances of the low compliance of residents and insufficient funds, sanitation units have to use limited funds to complete the most basic clean-up work so as to maintain normal operation, which affects the implementation of classified collection and transportation of waste.

Under the interaction between the contradictions in rights and responsibilities and funds, waste sorting has been in a "tolerant" state in recent years. The ineffective governance mode of "mixed collection and transportation, and combination after classification" has been continuously questioned by the public.

4. Incoherent policies

Shanghai is one of the earliest Chinese cities to implement pilot waste sorting. However, its specific implementation rules for waste sorting have been altered over and over again, affecting the implementation effect of policies.

Since 1999, Shanghai's waste sorting has undergone a series of adjustment and adaptation processes, where the specific standards and details for household waste sorting, namely the technical terms of categories of household waste, have been significantly replaced and adjusted. This has not only facilitated waste sorting, but also brought a challenge to the receptivity of target groups including citizens and units. The municipal government and environmental protection authorities have publicized the latest waste sorting standard by multiple channels concerning the press and media to enhance citizens' awareness of household waste sorting to some extent, but the ever-changing classification methods have made it difficult for citizens to fully understand or adapt to the specific rules. This has affected not only their enthusiasm for indoor waste sorting, but also the accuracy of community waste sorting.

5. Social participation to be improved

According to the questionnaire survey on pilot housing estates carried out by relevant units, the pilot work has achieved certain results, but there are still a high percentage (a quarter) of residents failing to take the initiative because of "shirking trouble" or "knowing nothing about how to classify waste". How to enhance these residents' awareness of waste sorting, encourage them to change their notion and take an active part in household waste governance has become an important part of breaking the bottleneck of waste governance in the implementation stage. Relevant departments should allow the publicity, education and guidance of waste sorting to approach communities and residents' lives in a deeper, more vivid and more specific fashion.

Survey findings also suggest that the implementation level of household waste sorting is closely related to age and pilot areas. From the perspective of age, residents aged 61–69 are the main force of household waste sorting; the young people are poor in participation, and only 8.4% of them have executed indoor waste sorting. From a regional perspective, different pilot areas vary in waste sorting level, and Chongming has achieved the most excellent results in the pilot work while Yangpu and Jinshan have achieved poor results. As regards the five indicators including "classification accuracy", "participation in the green account", "setting of dustbins", "residents' classification habit (including two parts, indoor classification and disposal)", as well as "evaluation on housing estate atmosphere", Chongming ranked among the top three, and topped the list in terms of four indicators, which indicated that it achieved significant results in the pilot work; Yangpu and Jinshan ranked among the bottom three on the whole, and thus had plenty of room for improvement.

Therefore, in the stage of fully advancing household waste sorting, how to improve the weak links of each district, and boost the overall coordinated development of the city's household waste governance has become the focus of the next stage.

9.4.2 Core of Household Waste Governance in Shanghai

1. Strengthening legal construction

In consideration of the problems in the practice of household waste governance in recent years, Shanghai developed and issued the *Implementation Plan for Establishing and Improving the Whole-Process Sorting System of Household Waste in Shanghai* (hereinafter referred to as the *Implementation Plan*) in 2018, where a series of clear and specific rules and regulations have become strong policy support and guarantee for propelling the municipal solid waste governance in Shanghai. The main principles of the *Implementation Plan* include: government promotion and public involvement; whole-process sorting and overall advancement; policy support and legal guarantee; urban–rural integration and measures appropriate to local conditions. Besides, the *Implementation Plan* sets strict time nodes and goals for the progress of municipal solid waste, striving to basically complete a whole-process sorting system of household waste in line with Shanghai's development orientation of being an excellent global city by the end of 2020. Moreover, in terms of resource recycling, valuation and appraisal mechanisms, etc., the *Implementation Plan* establishes a series of specific goals, including the following goals.

(1) Basically realizing the mandatory sorting of unit household waste in the city, and widely implementing the household waste sorting system in housing estates;
(2) Basically achieving "Dual-Network Integration", with the resource recycling rate of household waste up to 35%, and "Reduction, Resource Utilization and Harmlessness" reaching the domestic leading level.
(3) Establishing the up-to-standard acceptance listing system for household waste, promoting the building of demonstration housing estates (villages) and subdistricts (towns) for waste sorting, and continuing to improve the sorting effect.

In 2018, Jing'an District, Changning District, Fengxian District, Songjiang District, Chongming District, and Pudong New Area (urban area) took the lead in widely implementing the household waste sorting system, and built three national demonstration districts for rural waste sorting, and 700 administrative villages for waste sorting demonstration. Additionally, the *Implementation Plan* defines the specific work focus, which is mainly manifested in five aspects concerning terminal disposal, classified transportation, disposal quality, classification standards and regulatory systems, covers all parts of municipal solid waste governance; and employs regulatory measures to ensure the orderly and efficient advancement of the work.

2. Enhancing terminal disposal capacity

In light of the insufficient waste disposal capacity of the city, the *Implementation Plan* proposes that the terminal disposal capacity of household waste needs to be improved from the aspects of hardware and software. On the basis of improving the overall plan for waste disposal facilities, it is advisable to optimize modes and processes by reference to domestic and overseas advanced system design so as to improve the efficiency of household waste sorting and resource recycling.

First, as regards the layout and construction of harmless disposal and resource utilization facilities of household waste, it is necessary to make a reasonable layout on the basis of combining the overall spatial planning and layout of the city, and determine the plant site to break the "NIMBY" (not in my backyard) dilemma. Moreover, in light of the ever-changing new situations and requirements for the city's household waste sorting, relevant authorities should work out a plan based on the principle of "overall planning for functions, rational layout, and land saving". Considering the fact that some high-density areas in the city proper lack sufficient land for construction, it is advisable to renovate the completed household waste disposal facilities, fully tap the spatial potential of surrounding areas, and fulfill the specialized planning goal of municipal solid waste by "reconstruction and new construction".

Second, as regards the effective improvement of household waste disposal efficiency, it is necessary to draw upon the existing governance experience from home and abroad to conduct the process and system design of waste clean-up. The *Implementation Plan* indicates that from the perspective of the overall framework, it is advisable to establish and improve a harmless disposal and resource utilization system for waste, form a "big cycle" of household waste within the city, a "medium cycle" within a district, and a "small cycle" within a town (township), and build a three-level sorting system with the aim of organic combinations and constructive interactions. On this basis, all districts should take the opportunity of the transformation and upgrading of industry parks to advance the construction of district-level venous industry parks, and transform the traditional economic model of "resources - products - waste" into a closed-loop economic model of "resources - products - renewable resources" based on a resource cycle. A two-step approach of "pilot work + promotion" should be employed so as to carry out circular economy throughout the city, and form the distribution pattern of venous industry parks with "One Focus and Multiple Points". In addition, Shanghai should actively advance the building of a citywide recyclable distribution center, make a layout for the city's renewable resource industry in combination with the construction of the recycling industry park on the basis of the national market, and elevate the city's resource recycling level from the perspective of industrial planning.

3. Strict execution of classified transportation

Considering the contradiction of "combination after classification" in the collection and transportation process of household waste, the *Implementation Plan* explicitly specifies the standardized collection and transportation process and regulatory mode of all types of household waste. Sanitation units and authorities must implement the mandatory classified collection and transportation of household waste in accordance with the standard, and strictly supervise and implement the system of sorting managers pursuant to the provisions of the *Measures of Shanghai Municipality for Promoting Household Waste Sorting and Reduction*. In case of "mixed loading and transportation" in the transportation process, relevant authorities must severely hold the sorting managers accountable. Additionally, in case that sorting quality does not meet the standard for classified collection and transportation, relevant management authorities should urge the sorting managers to organize secondary sorting so as to ensure the sorting quality of household waste.

For the collection and transportation of household waste, sanitation collection and transportation enterprises should strictly execute the classified collection and transportation norm, implement collect and transport all sorts of household waste in accordance with the sorting standard, resolutely forbid mixed loading and transportation, and accept the supervision from all sectors of society by means of publicizing the collection and transportation time, and standardizing vehicle types and logos. Management authorities should enhance daily supervision and assessment, bring the implementation status of classified transfer, collection and transportation norms of household waste into the assessment and review systems for relevant enterprises; and seriously investigate and punish any sanitation collection and transportation enterprises for mixed loading and transportation. Moreover, pursuant to the requirements raised by relevant management authorities, the aforesaid enterprises should establish a mechanism of "requesting rectifications for the first violation, collecting and transporting waste upon rectifications for the second violation; rejecting collection and transportation, and transferring the matter to law enforcement authorities for punishment in case any units refuse to make rectifications after multiple violations" for any units failing to earnestly implement waste sorting or failing to meet the sorting requirements, and urge units to strictly implement waste sorting by means of the mandatory measure of "no sorting, no collection and transportation".

4. Improving waste dumping quality

Considering the city's low-quality waste dumping and unsatisfactory sorting effect, the *Implementation Plan* proposes making improvements in four aspects, including dumping at a "fixed time and place", deepening the green account incentive mechanism, accelerating the "Dual-Network Integration" process of household waste sorting and renewable resource recycling, and improving the system of "separation for bulky waste".

First, the implementation of dumping "at a fixed time and place" should be combined with the "Beautiful Home" campaign under construction by housing estates throughout the city. With the classified dumping of residual and household food waste as the main content, the sorting site setting, dumping time arrangements and sorting norms for dumping "at a fixed time and place" should be determined according to the actual conditions of housing estates. Within housing estates, property management companies, neighborhood committees and other sorting managers should perform the publicity, guidance, as well as daily supervision and coordination with respect to residents; for commercial tenants along streets, relevant operators should actively cooperate with the working system of "door-to-door collection" by sanitation departments, and deliver sorted waste at a fixed time and place.

Second, the further improvement of the green account mechanism should be oriented towards "Voluntary Application, Self-Service Point Accumulation, and Free Exchange" to extend the opening channels of the green account, and to continue to expand the coverage thereof. The *Implementation Plan* explicitly proposes striving to achieve the simultaneous coverage of the green account and waste sorting by 2020. In the implementation process of waste sorting, the government should work with enterprises and social organizations to improve the point rules of the green account, and give play to the incentive role of the green account in stimulating waste sorting and recycling; enhance government procurement, and introduce enterprises and other third parties to the publicity, guidance and supervision of waste sorting so as to explore a market-oriented operation mode for the green account; employ the "government-led market participation" mode to raise resources by multiple channels, enhance the influence of the green account on residents and its attraction for them, and give full play to its positive incentive effect.

Third, based on the principle of "separation with combination, classification in stages", landscaping, city appearance and sanitation authorities, as well as commercial departments should untangle the management responsibilities for renewable resource recycling, and speed up efforts to propel the "Dual-Network Integration" of the renewable resource recycling system and the classified collection and transportation system of household waste. All departments should actively cooperate with the integration of classified dumping sites and delivery sites of the "two networks", optimize and improve the renewable resource recycling function of dustbin rooms; advance the participation of enterprises and public institutions in the "Dual-Network Integration"; implement the "One Position with Two Duties" training for renewable resource recycling personnel and household waste sorting personnel to gradually achieve the unified management of "two networks"; seek and cultivate recyclable collection and transportation enterprises in all districts to put the addition of renewable resource recycling business to sanitation enterprises into practice; arouse the enthusiasm of enterprises and social organizations, and encourage them to establish second-hand goods adjustment platforms, junk shops, flea markets, and other waste goods trading platforms so as to promote the reutilization of idle resources.

Fourth, from the perspective of the overall process of waste governance, the city should insist on the basic path of "separation for bulky waste, and classification for small ones", and improve the system of "separation for bulky waste". On the one hand, it is necessary to strengthen decoration waste management, standardize the setting of dumping sites in housing estates, guide residents to classify decoration waste from the source and stack it in bags; and enhance the collection, transportation and recycling services of bulky waste. On the other hand, it is advisable to promote the separate collection and utilization process of dry branches and fallen leaves, and encourage landscaping and maintenance enterprises to conduct the resource utilization thereof by means of smashing and composting. Moreover, the city should improve the waste separation and classification system in peddlers' markets, and combine the construction of household food waste collection and transportation and resource utilization systems to elevate the waste resource utilization level thereof.

5. Clear standard for waste sorting

Since the 1970s, Shanghai has changed the technical terms of waste sorting for many times. It was not only a huge challenge to citizens' receptivity, but also brought much inconvenience to waste sorting. Therefore, since the launch of a new round of household waste sorting in 2011, Shanghai has been insisting on the classification basis of "residual and household food waste", and implementing the four-category sorting standard. The latest *Implementation Plan* further proposes that Shanghai will continue to implement the four-category standard of "hazardous waste, recyclables, household food waste, and residual waste", which not only conforms to the requirements of documents issued by the General Office of the State Council, but also helps to consolidate and extend the sorting results in the early stage. Besides, the *Implementation Plan* encourages all units and housing estates to classify household waste in a more detailed way based on their resource utilization capacity so as to reduce waste output at the source.

Additionally, the city will standardize the setting of household waste sorting containers, and allocate sorting managers so as to facilitate the correct classification of waste by residents and improve the efficiency of waste governance. As regards housing estates, units and public places, different waste collection facilities should be arranged according to their own characteristics; and waste should be collected separately based on the principle of "convenience and controllability".

6. Strengthening the policy supervision system

Considering the problems like unclear rights and responsibilities and inadequate implementation of relevant management authorities exposed in the prior practice process, the *Implementation Plan* defines the work scope, duties and objectives to be fulfilled by respective departments. During daily work, all departments should not only perform well in notifications regarding classification obligations, classification standards, and responsibilities for classified disposal management from the perspective of flexible guidance, and provide business training for related managers, but also strengthen rigid supervision, timely stop any violations of waste sorting regulations, and strictly urge rectifications; take coercive means of administrative law enforcement in case any units or individuals refuse to implement waste sorting, and transfer them to the urban management law enforcement department for punishment. In addition, the urban management law enforcement department should strengthen daily inspection and law enforcement against any violations of waste sorting pursuant to the provisions of laws and regulations, and punish any units and individuals for refusing to perform the sorting obligation upon critical education in accordance with laws and regulations.

Additionally, to further urge government authorities at all levels to effectively execute the implementation rules for waste sorting, the *Implementation Plan* specifies the working platform of the two-level (city and district levels) joint conference for household waste sorting, and brings the waste sorting governance strength and effect into the system of administrative performance assessment and assessment on ecological civilization performance indicators. It is aimed at urging and encouraging all district governments to attach importance to the orderly advancement of household waste governance by improving and optimizing the assessment system. From the perspective of strengthening legal guarantee, improving the policy system and implementing inspection mechanism, it requires reinforcing local legislation and cooperating with several departments in administrative law enforcement; improving the subsidy policy for local household waste disposal facility construction and the environmental compensation system for cross-district disposal, optimizing the support policies concerning energy conservation and emission reduction that promotes source classification and reduction; intensifying and establishing a whole-process integrated supervision system for household waste sorting characterized by government supervision, professional third-party regulation, and public involvement.

Furthermore, to actively advocate the household waste sorting system and enable the green lifestyle to win support among citizens, the government should optimize its publicity, education and assessment system, and enhance citizens' environmental awareness. It is advisable to strengthen the community publicity and door-to-door publicity of waste sorting by attaching importance to the publicity guidance of sub-districts and neighborhood committees; bring waste sorting into the preschool and compulsory education curriculum system, and build a number of demonstration education bases for household waste sorting to foster residents' awareness of waste sorting from childhood by means of teaching through lively activities, and to guide them to establish a concept of green life; increase the percentage of waste sorting in the civilization building assessment system, and arouse units and individuals' enthusiasm for participating in waste sorting by combining rewards and punishments.

Chapter 10
Effective Supply Mechanism Design of Environmental Public Goods

In consideration of the "government failure", "market failure" and "voluntary failure" in different supply modes of environmental public goods, through analyzing environmental pollution sources, this book affirms the effectiveness of the government's "Pigovian tax" policy targeting enterprises on the one hand, and identifies the existence of Pareto improvement on the other hand; fully demonstrates that the government-led mode of motivating residents and enterprises to participate in environmental governance is capable of bringing the effect of Pareto improvement through theoretical and empirical research, and proposes the necessity to highlight the guarantee mechanism of intergenerational contracts and guide residents to participate in environmental governance, with a focus on how to motivate residents to participate in environmental governance as the supply side.

10.1 Logical Path of Motivating Residents to Serve as the Supply Side of Environmental Public Goods

The preceding part has expounded such issues as China's environmental status, residents' role in environmental supply, changes in material life conditions, residents' WTP for environmental protection. However, to motivate residents to serve as the supply side of quality environment, mechanism design appears particularly important.

Through analyzing the realities of some typical environmental public goods, including air, water and waste pollution, this book intensively reflects the fact that China's supply of environmental public goods is severe and dominated by the single mode of government supply, and residents, as one of the pollution sources, have been only acting as the demand side of quality environment, and fail to be motivated to serve as the supply side thereof. The main reason lies in the "free-rider" problem of public goods, which results in the impossibility of revealing residents'

J. Yang, *Effective Supply Mechanism and Path of Environmental Public Goods in China*, Public Economy and Urban Governance in China,
https://doi.org/10.1007/978-981-99-2939-9_10

real demand for environmental public goods. If it is possible to embody residents' effective demand for environmental public goods by mechanism design, motivate residents to reveal their real demand, and consider them as the supply side, the effective supply of environmental public goods will be certainly improved.

Question 1: Is it possible for residents to serve as the supply side of quality environment?

According to the theory of economic development presented by Rostow, in case of entering the stage of the "search for quality", the people will be more concerned about environmental problems, and more willing to participate in environmental protection. Through analyzing per-capita GDP and Engel's coefficient, this book points out that Chinese residents' material life conditions have been capable of reaching the stage of the search for quality, and laid a foundation for residents' participation in environmental protection.

Analyzing residents' material life conditions, the above not only answers Question 1: it is possible for residents to serve as the supply side of quality environment, but also brings Question 2.

Question 2: By what approach can residents realize the shift from the demand side to the supply side of quality environment?

CGSS data suggests that over half of Chinese residents are willing to pay fees or taxes for environmental protection; if neutrals are added, the percentage will rise to more than 60%. The empirical survey not only provides strong support for us to motivate residents to participate in environmental protection, but also can be used as evidence for the answer to Question 1. It shows that residents are willing to pay for environmental protection, and answers Question 2: residents can realize the shift to the supply side of quality environment by making a payment to some extent. CGSS reflects residents' real preference for environmental public goods on some level. Next, we will further probe into Question 3.

Question 3: How do we put residents' WTP for environmental protection into practice? How do we effectively measure residents' WTP for environmental protection?

To motivate residents to reveal their real demand and solve the "pricing difficulty of public goods", residents' WTP for environmental public goods has been discussed in the academic circles. This book compares some typical methods, including the revealed preference method, the stated preference method and the happiness measurement method, holds that the happiness measurement method is more scientific than other methods, and intensively sets forth the idea, calculation and standard setting of the happiness measurement method in measuring residents' WTP for environmental protection. This methodology provides a scientific path for motivating residents to participate in environmental protection, and offers a solution idea for Question 3, i.e. residents' WTP for environmental protection is measurable.

Question 4: How will environmental protection fees be regulated, and by whom?

Due to the difference in income levels, per-capita GDP varies in different places, and the government is required to set the environmental protection payment standards for residents in different places; besides, in the event that material conditions support residents' pursuit of quality environment, or empirical survey data suggests residents' willingness to pay for environmental protection, the government needs to enact policies for stimulation as an institutional guarantee. This is another core issue of this book.

Question 5: How do we design an effective mechanism to guarantee the effective supply of environmental public goods?

The foregoing Questions 1 to 3 are all expounded in previous chapters, thus the following part will focus on Questions 4 and 5 to intensively explore the effective supply mechanism of environmental public goods.

10.2 Theoretical Basis and Framework for Supply Mechanism Design of Environmental Public Goods

In existing research, the mechanism issue is explored from the perspective of egoistic and altruistic motivations. Premised on the assumption of an egoistic motivation, Doeleman and Sandler [75] built a finite OLG model, researched future generations' investment in intergenerational public goods, and concluded that underinvestment was inevitable as every generation was self-interested. Starting from an assumption of traditional economics, if it is believed that humans are egoistic in nature, middle-aged people are more likely to invest in backward intergenerational public goods like old-age security, but unwilling to invest in forward intergenerational public goods, e.g. offspring education, and environmental protection. However, from the perspective of China's national conditions and families extended across generations, the vast majority of families will not neglect offspring education, health or other investment, because on the one hand, from an emotional perspective, they consider that they have a great responsibility to cultivate offspring—"holding high hopes for their children"; on the other hand, they hope to gain returns after the success in offspring cultivation—"raising sons to provide for their old age", and believe that more successful offspring education will bring them higher returns. In the view of Becker [170], the role of altruism in family life is as significant as that of selfishness in market transactions.

Based on this circumstance, we assume that present generations are neither totally altruists, nor totally egoists, and they tend to not only attach importance to the education and future development of their offspring, but also have some self-interested considerations, and may plan for their future old-age support [171]. Therefore, in terms of residents, out of an altruistic motivation, they are willing to pay for environmental governance; out of an egoistic motivation, they also hope that the government could reciprocate to some extent.

10.2.1 Theoretical Basis

Residents' voluntary payment is conditional. Because of not knowing the government's real cost for pollution control, and being not sure whether the government will fulfill its commitment, residents form principal-agent relations with the government. Meanwhile, to avoid "Government Failure" (e.g. the government may show a semblance of cooperation in accepting residents' voluntary payment for environmental protection, reveal the high cost of environmental governance to the greatest extent possible, or fail to act), an "environmental protection fund company" in market competition must be established to operate the environmental protection fund, and the environmental governance performance of the fund company must be supervised by residents. In this way, the government and the "fund company" can establish principal-agent relations as well. Due to the incompleteness of contracts made between residents and the government, and between the government and the fund company arising from information asymmetry, high anticipated costs and contractual costs before payment, and high confirmed costs thereafter, the government-led supply mechanism design of environmental public goods is an incomplete contractual behavior.

1. Principal-agent theory and incomplete contracts

Principal-agent relations refer to the relations of responsibilities, powers and interests between the principal and the agent formed by an arrangement in which the principal appoints the agent to act on its behalf. The principal-agent theory was first presented by American scholars Berle and Means [172] upon the research on a number of modern corporations. They found that the separation of ownership and control (power of operation) occurred in many corporations, which were actually controlled by the "controllers group" composed of professional managers; principal-agent relations appeared with the gradual separation of corporate ownership and control. The core task of the principal-agent theory is to research how the principal determines the optimal contract incentive agent in the context of the information asymmetry and the conflict of interest. It follows that the principal-agent theory is premised on the inconsistency of interests and information asymmetry between the principal and the agent.

The suppliers of environmental public goods are under the circumstance of information asymmetry. With an urgent desire for pollution control, residents voluntarily pay the environmental protection fee, but obviously the government has an information advantage when using this fund. The moral hazard may lead to the inconsistency of interests between them. When the government entrusts the task of environmental governance to the environmental protection fund company, the government becomes the weak party, and under an egoistic motivation, the fund company is likely to do something to the detriment of the government (precisely, public interests). Thus, it is necessary to sort out the relations between the parties in the process of mechanism design.

2. Problems in incomplete contracts and solutions thereto

Due to the inconsistency of interests between both parties to a principal-agent contract, agency cost may easily arise. In the view of Jensen and Meckling, the principal cannot fully motivate the agent, and the principal's cost for monitoring the agent is probably higher than revenue, thus it is impossible to implement perfect monitoring on the agent [173]. Arrow [174] divided the principal-agent problem into moral hazard and adverse selection. Moral hazard refers to an action taken against the principal by the agent, who knows that the principal has difficulty in observation and monitoring. Adverse selection means that the agent possesses information that the principal cannot observe, and use such private information to make decisions.

To solve the problems in principal-agent relations, scholars conducted continuous explorations, and successfully proposed the agency model, the reputation model, the effect model, the monitoring model, the selection model, and so forth. Using a repeated game model, Radner [175] demonstrated that if the discount factor was big enough, i.e. the principal and the agent had sufficient confidence and maintained long-term relations, the Pareto first-order optimal risk sharing and incentive could be achieved. Fama [176] explicitly put forward the reputation problem, and held that even without an explicit incentive contract the manager (agent) might still work actively, because this could improve its reputation in the managerial market, and improve future income. Holmstrom and Ricart-Costa [177] researched the "ratchet effect", similar to the "bullwhip" phenomenon that we usually mention. It means that the people working harder can bring higher output, which causes enterprises to set a higher standard in the next year, and results in punishment contrarily; thus, those "clever" people conceal their production capacity to cope with the planning authorities. Upon research, Holmstrom and Costa held that the ratchet effect would weaken the incentive mechanism in a long-term process. Through model analysis, McAfee and McMillan [178] considered not only the moral hazard in teamwork, but also the adverse selection therein, and proposed that the solution to the principal's moral hazard was allowing the principal to monitor the agent, rather than collect a deposit from the agent. Solow [179], Shapiro and Stiglitz [180] researched the inevitable monitoring problem in principal-agent relations. They held that high wage could be used as an incentive method to prevent workers from shirking; higher wages would lead to larger opportunity costs of being unemployed. Zhang and Yu [181, 182] put forward a more basic question in principal-agent relations: Who should be the principal, and who should be the agent in a particular organization? Or how should principal-ship (power of entrustment) be allocated to different members? His research conclusion is that the determinant of the optimal assignment of principal-ship is the relative importance of enterprise members to production, and their relative effectiveness to supervision. Another solution to the principal-agent problem is allowing potential agents to compete with each other, thereby forming a mutually restrictive mechanism between agents. In the view of Pindyck and Rubinfeld [183], an incentive mechanism in the principal-agent framework could be built. They analyzed the design of a profit-sharing arrangement and bonus payment system to solve all principal-agent problems, and considered that in case it was impossible to directly measure the results of efforts, the incentive structure that gave rewards to high-level effort results could enable the agent to pursue the goal set by the owner.

In the supply process of environmental public goods, the principal-agent relations between residents and the government, and between the government and the environmental protection fund company, will be inevitably faced with the above problems, i.e. residents' supervision on the government, the government incentives and supervision for the environmental protection fund company, as well as the basic problem: who is more reasonable, the agent or the principal? It shows that how to design an incentive mechanism has become one of the core parts for effective supply of environmental public goods.

10.2.2 Mechanism Design Framework

Based on the above theoretical analysis, in the process of mechanism design, we should take residents, enterprises, the government and the environmental protection fund company as the behavior subject (actor) for analysis, extend the traditional government supply mode that is implemented by collection of the Pigovian tax to public participation in environmental governance, and allow all parties to perform their own functions, make respective contributions, and march towards the common vision of blue skies, white clouds, lucid waters and lush mountains.

1. Design of a mechanism where residents and enterprises jointly participate in environmental protection led by the government

We describe the supply mode of environmental governance with the joint participation of the government, enterprises and residents as the mode of government-led voluntary supply by residents and enterprises. This mode is different from the traditional government supply mode. For the latter, the government is the sole supply subject; for the former, the supply subject is the entire people (the government, enterprises and residents). The government plays utterly different roles in the two supply modes. In the mode of government-led voluntary supply by residents presented herein, the government plays a guiding and motivating role, and is responsible for system design guarantee, as well as the intergenerational equity commitment to residents and the fulfillment thereof (equity-based incomplete contract design).

As regards the mechanism design for the mode of government-led voluntary supply by residents, the most crucial problem lies in how to motivate the entire people to participate therein?

Take household waste disposal as an example. The design idea of the mechanism where residents and enterprises jointly participate in environmental protection led by the government is described in Fig. 10.1.

The design of the above mechanism covers the government, enterprises, residents, as well as an environmental protection fund company, the community sorting and

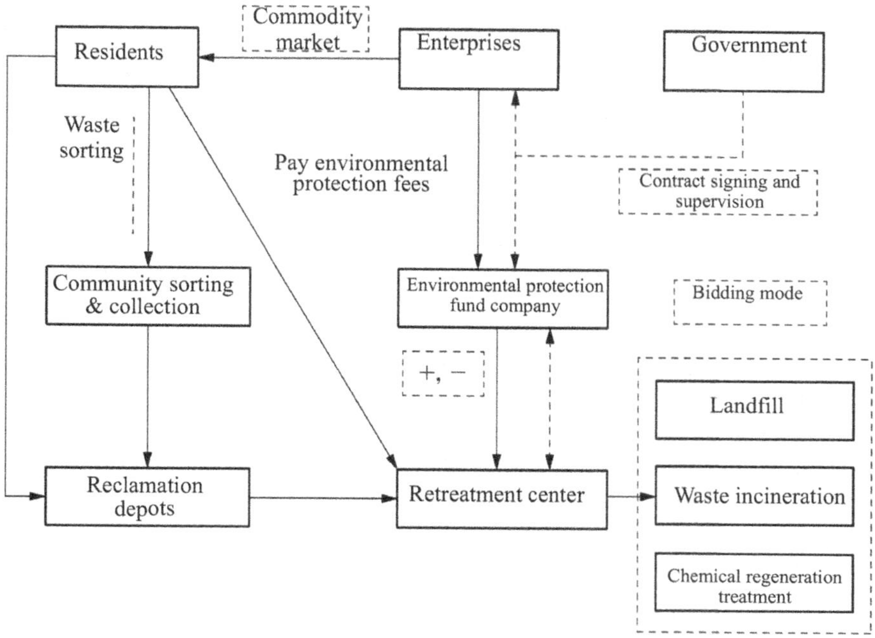

Fig. 10.1 Design idea of the mechanism where residents and enterprises jointly participate in environmental protection led by the government

collection department, reclamation depots, a retreatment center, and so on. Their relations of responsibilities, powers and interests are described as follows.

(1) Government

As the main undertaker of environmental protection, the government plays a big role in the mechanism design. It is responsible for the entire mechanism design and the operation framework, and plays a role in supervising environmental protection on the whole; more importantly, it signs a contract with enterprises and an environmental protection fund company respectively, collects pollution fees from enterprises to force them to lower pollution, and selects a fund company by means of market competition, which is mainly responsible for environmental protection operation to achieve the basic goal.

(2) Residents

Residents have always been in the "free-riding" state. But based on the preliminary analysis on income, environmental awareness and environmental action, the government can motivate residents to participate: On the one hand, learning from Japan's original idea of waste sorting, and implementing different waste charges according to the size of garbage bags to motivate residents to reduce waste output; on the other hand, giving collection fees for recyclable waste, including paper waste, plastic waste, and waste glass. In other words, residents have to pay for throwing household waste, and may get paid for collecting "good" waste.

(3) Enterprises

Polluting enterprises (polluters) should be responsible for the entire society. When they turn on machines to produce products, it means that pollution arises. Thus, for the mechanism design, on the one hand, enterprises should provide residents products needed by society; on the other hand, they should pay environmental protection fees to the environmental protection fund company to assume their responsibility.

(4) Environmental protection fund company

Similar to a platform, the environmental protection fund company is selected by the government by means of market competition. It is responsible for managing environmental protection fees, controlling waste pollution, signing a contract with the government, and selecting a waste retreatment center by bidding. If the retreatment center disposes waste properly and can create benefits for society, it will get rewards, denoted by "+"; if it does not dispose waste properly, or fails to achieve the anticipated goal, it has to pay fees to the fund company, denoted by "–".

(5) Retreatment center

The waste retreatment center is mainly responsible for disposing household waste, and must scientifically verify what method is the most conducive to the public. Methods include landfill, incineration, and chemical regeneration treatment. The center is also a result of market-oriented operation, and selected by the environmental protection fund company through public bidding.

(6) Community sorting and collection department

The department is proposed based on China's current situation, equivalent to community neighborhood committees or other similar management departments. After residents classify waste, the most convenient way for them is to throw waste in the waste dumping site of their housing estate, and pay the waste treatment fee. The community sorting and collection department must properly provide waste sorting guidance (learning from German or Japanese mode), and correctly classify waste.

(7) Reclamation depot

Considering the recycling value of some waste, after waste sorting and collection in communities, secondary sorting may be implemented, and different types of waste may be sold to the retreatment center.

The above mechanism design can motivate residents to reduce waste pollution, and promote the operation of the green and environment-friendly recycling system by requiring residents to pay for waste treatment and allowing them to get paid from the sales of recyclable waste. This supply mode of environmental public goods that motivates residents to participate has the effect of Pareto improvement. However, it has a drawback. Taking the means of collecting pollution fees alone to cope with the pollution caused by enterprises, the mechanism fails to form a closed loop.

In other words, this mechanism will not spur enterprises to reduce pollution by vigorously researching and developing technology unless the punishment intensity, namely the environmental protection fee, is fairly high (some scholars have employed the cost-benefit approach to assess this method, and concluded that the disadvantage outweighs the advantage on the long run).

Well, how do we design a more effective mechanism to effectively stimulate the environmental protection behavior of enterprises and residents?

2. Design of a mechanism where residents and enterprises jointly participate in environmental protection led by the government from an intergenerational perspective

Considering the advantages and disadvantages of mechanism design, we are working on how to feed the subsequent results back to enterprises and residents, enable the two social subjects to actively participate in environmental governance, and form an effective supply mechanism of environmental public goods.

How will the "free-riding" behavior be avoided in mechanism design? What kind of supervision is effective in principal-agent relations? How will moral hazard and adverse selection be lowered in the design of an incomplete contract? After making explorations with these questions in combination with previous theoretical and empirical analysis results, the author held that as pollution sources, enterprises and residents should be discussed separately. Pigovian taxes should be imposed on enterprises. For enterprises failing to meet national standards, they should pay taxes or be regulated; but if some enterprises have updated technology, and are capable of bringing revenue (with positive externality) to relevant industries and the society, they should receive corresponding incentives. As regards residents, on the one hand, they may produce such pollution as air pollution (e.g. traveling by vehicle), waste pollution and water pollution in daily life. Because of the difficulty in obtaining their real demand for environmental public goods, residents' "free-riding" behavior is widespread. The government should fully survey residents' WTP for environmental protection, influencing factors, etc., design appropriate incentive policies to encourage those residents who are willing to pay for environmental protection on their own initiative, and take into account their egoistic and altruistic motivations to safeguard their interests. On the other hand, the government should introduce measures to guide and encourage those residents who try to get a "free ride" to go into action, fully consider their utility function and objective, and motivate them to devote themselves to environmental protection from an egoistic motivation perspective.

Based on the above considerations, this book presents the following mechanism design and policy research idea as shown in Fig. 10.2.

Transfer the environmental protection fund and discount to residents in period (t + 1).

In Fig. 10.2, there are four subjects, i.e. the government, enterprises, residents, and the environmental protection fund company. To further achieve the effect supply of

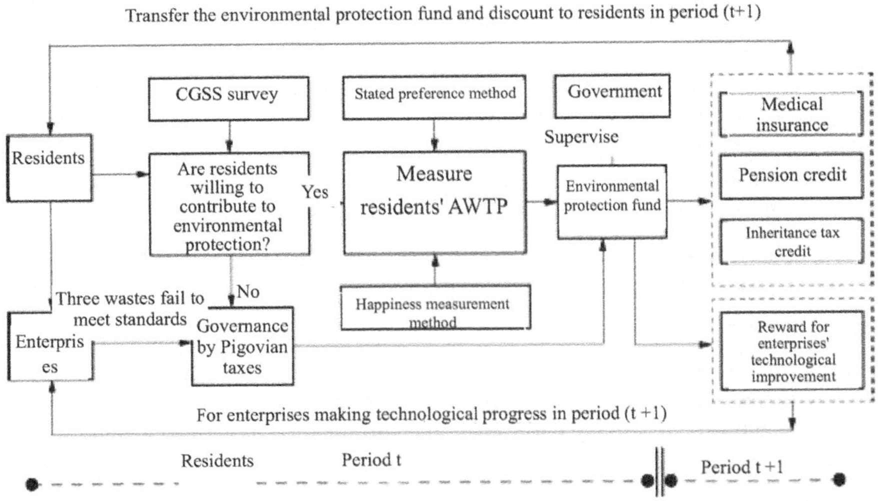

Fig. 10.2 Design of a mechanism where residents and enterprises jointly participate in environmental protection led by the government from an intergenerational perspective and policy research idea thereof

environmental public goods, the four subjects perform their own functions according to respective characteristics and objectives.

(1) Government

In the environmental protection mechanism design, the government is the most important subject, which has exceptional advantages in terms of both the formulation of macro policies and the supervision of the micro environment. However, if the government is in full control, it may easily cause insufficient participation of social forces, and no room for market opportunities, so the boundary setting of government participation is quite critical. In the mechanism design herein, the government is mainly responsible for developing policies and measures, and selecting and supervising the environmental protection fund. On the one hand, the government regulates the environmental pollution arising from economic activities; on the other hand, it can further motivate enterprises and residents to actively join the "supply side" of quality environment on the basis of imposing taxes on polluters. Given that environmental governance is not only a spatial externality issue, but also an inter-temporal intergenerational issue, in the supply process of intergenerational public goods, the government can design an "intergenerational contract". The details are described as follows.

As regards residents, according to the data of CGSS, over half of residents are willing to pay fees or taxes for quality environment. The government can estimate the payment benchmark by the happiness measurement method, and introduce policies to set an "environmental protection fund". Residents who are willing to pay in the middle-age period (period t) can receive subsidies from the government in the old-age period (period $t + 1$) in the forms of old-age insurance increase, medical insurance increase, inheritance tax credit, etc. This "intergenerational contract" is capable of motivating residents to actively participate in environmental governance. Certainly, there is no compulsion on residents who are unwilling to pay. The governance plan of "Pigovian taxes" can be adopted. However, residents who are unwilling to pay cannot get corresponding benefits in the old-age period (period $t + 1$).

The government policies targeting enterprises should not only encourage enterprises to "internalize" their spatial externality problem, but also motivate them to make technological improvements. This can not only improve environmental quality, but also promote the capital accumulation of enterprises and boost economic growth. Specifically, in case that the pollutant discharge of enterprises fails to meet the standard (in period t), the pollution tax policy can be used to impose taxes on enterprises, and the tax revenue will enter the "environmental protection fund" pool. In the event that enterprises have made technological improvements to the benefit of environmental protection, they can get rewards from the government in period $t + 1$. The rewards do not have to be material incentives, and may be preferential policies, e.g. precedence in bank loans, or government guarantee for credit.

(2) Enterprises

The enterprises referred to herein mean enterprises causing pollution. The government's environmental regulation will definitely reduce the profits of enterprises. Thus, from the perspective of enterprises, they may attempt to seek rents from competent authorities, or directly make a countermove, i.e. "devising countermeasures against government policies". In any case, government policies must ensure that "there are laws to go by, the laws are observed and strictly enforced, and law-breakers are prosecuted"; the environmental protection fund company and residents jointly conduct supervision; enterprises must undertake social responsibility, and pay pollution control fees pursuant to the environmental protection policies and systems developed by the government. Furthermore, to encourage enterprises to make continuous technological improvements, upgrade production capacity and lower pollutant discharge, the government should take incentive measures to encourage enterprises to realize energy conservation and emission reduction by making technological progress while strictly curbing pollution. The combination of punishments and rewards can help enterprises to continue to pursue better technology, and achieve more effective pollution control.

(3) Residents

As the sufferers of pollution, residents are most eager to see the increasing improvement of their living environment. First, to avoid the "free-riding" behavior, the mechanism design herein distinguishes between residents paying for environmental protection and those inactive ones (namely, free-riders). Residents who voluntarily pay for environmental protection in the middle-age period can receive subsidies in their old age in the form of medical insurance or pension increase. As regards the amount of subsidies, the discount rate should be taken into full account. In other words, it is necessary to ensure that residents who voluntarily paid for environmental protection could be properly paid in their old age, from the perspective of the economic man's egoistic motivation. Meanwhile, we suggest raising these residents' credit score in the credit system. Certainly, intergenerational enjoyment measures may be taken. Out of an altruistic motivation, these subsidies may be enjoyed by the offspring of those residents paying for environmental protection (considering accidental death or death from illnesses).

Second, considering that residents' behavior may also result in air and water pollution, local governments should enhance the publicity of green travel, and work with the environmental protection fund company to accomplish wastewater treatment and other treatment work. Only when residents' environmental awareness is enhanced will they choose to travel by public transport, and dump waste upon classification on their own initiative.

At last, residents has the right to oversee government actions, the environmental protection fund company, polluting enterprises, and so on, make complaints or put forward suggestions to relevant government authorities at any time.

(4) Environmental protection fund company

The environmental protection fund company is mainly responsible for applying the environmental protection fund paid by enterprises and residents into environmental governance in the most efficient way to ensure the effective operation of the fund. The company must be selected in a competitive mechanism so as to ensure the operation efficiency of the company. It is regulated by the government, and monitored by residents. The government may select competitive environmental protection fund companies according to different regions and types of pollution.

Certainly, the fund company is also required to undertake the responsibility for giving residents and enterprises intergenerational subsidies. Given that residents voluntarily pay pollution control fees out of egoistic and altruistic motivations in the young and middle-age period, they should enjoy subsidies, and the fund company should perform the previously signed contract. For enterprises, if they have improved technology, and transformed negative externality into positive externality, they should be rewarded.

Generally speaking, the effective supply mechanism of environmental public goods requires the government, residents, enterprises, as well as the fund company commissioned by the government to make a concerted effort. The government should introduce the overall design idea and the "long-term contract" performance guarantee for systems and policies to ensure that residents' intergenerational contract could be thoroughly implemented, and select a fund company to ensure the effective operation of the fund and fully achieve the government's policy objectives; residents should continue to improve their environmental awareness, and motivate more residents to reveal their real demand for environmental public goods and to voluntarily pay for environmental protection with the incentive of government policies; enterprises should bear the cost of pollution arising from them. Moreover, stimulated by the government rewards for new technologies, polluting enterprises will weigh the cost of pollution control and that of technological improvement, and apply new technologies to green production and operation.

It follows that to achieve the functioning of the mode for government-led voluntary supply of environmental public goods by residents and enterprises from an intergenerational perspective, supporting policies and measures must be equipped. First, the government's guarantee mechanism should be in place, and the government must fulfill its commitment; second, policies should play a role in motivating residents to participate in environmental protection.

10.3 Measures for Residents' Participation in Effective Supply of Environmental Public Goods from the Perspective of Intergenerational Externality

In recent years, China has continued to intensify efforts to control all sorts of pollution, e.g. controlling the total discharge of major water pollutants in the "Ninth Five-Year Plan"; proposing and completing the goal of reducing the emissions of COD and SO_2 by 10% respectively on the basis of that of 2005 in the "11th Five-Year Plan"; proposing the total emission control objective of SO_2 and COD in the "12th Five-Year Plan". The *Notice of the State Council on Issuing the Action Plan for Prevention and Control of Water Pollution* issued in 2015 proposes the following objectives: By 2020, water environment quality nationwide will be periodically improved, with heavily polluted water bodies dramatically reduced; by 2030, it is expected that overall water environment quality will be better and water ecosystem functions will preliminarily recover. The *13th Five-Year Plan for the Protection of Ecological Environment* clearly proposes that the overall objective is to improve the environmental quality by 2020; determines seven major tasks, including to combat air, water and soil pollution; and stresses that "lucid waters and lush mountains are invaluable assets", and "we should protect the ecological environment like protecting our eyes, and treat the environment like it is our lives".

To achieve these objectives, this book presents the mode for government-led voluntary supply of environmental public goods by residents and enterprises. Compared to the existing supply mode, this mode not only requires polluting enterprises to take social responsibility, and motivate residents, as the demand side with the most pressing need for clean environment, to assume corresponding social responsibility, thereby forming a three-sector (the government, enterprises and residents) cooperative governance mode. To achieve these objectives, the government must have a corresponding policy guarantee mechanism.

10.3.1 Abandonment of Short-Sightedness from the Perspective of a Long-Term Government[1]

According to the research findings of Shi et al. [184], environmental regulation plays a significant role in promoting urban economic growth, and the role gradually increases with the passage of the implementation time thereof, and has a progressive "increase of marginal utility" in terms of city size, i.e. environmental regulation plays a stronger role in promoting economic growth in larger cities. Therefore, when developing corresponding environmental protection policies, the government should consider long-term interests, abandon short-sightedness, properly handle the relations between long-term interests and short-term contradictions, and consider problems from the perspective of a long-term government.

Due to the externality problem of environmental public goods, there is not only the "free-riding" behavior of present generations, but also the absence of beneficiaries (future generations) in the process of environmental governance. The combination of spatial and temporal externalities deepens the difficulty in environmental governance. If the government still makes decisions on intergenerational public goods according to the duration of the term of office, the short-sighted phenomenon will surely emerge. For example, the government may tend to pursue projects which are capable of bringing profits within a short term, and only focus on economic growth and other projects which can easily bring political achievements. All these are not conducive to environmental governance. The long-term government is relative to a five-year term of office. It mainly means that the government should take a long view when considering problems, and give consideration to sustainable development, without prejudice to the survival benefits of future generations. Only when the government takes a long view can it is possible to solve the intergenerational externality problem of environmental public goods.

However, affected by fiscal decentralization and vertical political management, China's local governments tend to "be responsible to superiors", rather than "be responsible to subordinates". In the view of Bardhan et al. [186, 187], due to the imperfect democracy and rule of law, the local governments of developing countries are more likely to be "captured" by interest groups; in particular, fiscal decentralization may lead to the disregard for ordinary residents' needs, the overestimate of public service cost, the breeding of corruption, etc. China's fiscal decentralization system has brought about the fierce competition among local governments, and turned them from a "helping hand" into a "grabbing hand" [188]. Moreover, fiscal decentralization has caused the distorted structure of public spending where local governments "put capital construction above public service" [189]. Additionally, the assessment mechanism for government officials tends to be "GDP-centered", which

---

[1] This is borrowed from the view of Kotlikoff et al. [185]. The long-term government is relative to the short-term government, and means that the government should take a long view, and take into account the interests of both present and future generations; and its design of policies and measures should not be only based on the GDP assessment indicators during the term of office, but give consideration to sustainable development.

directly causes local governments to take no account of the supply of environmental public goods, because such achievements do not add weight to their promotion.

Considering the existing problems in combination with the mechanism for government-led voluntary supply of environmental public goods by residents proposed herein, we suggest the central government improve the assessment mechanism for local governments, especially remove the previous "GDP-centered" evaluation indicators, bring the environmental damage indicator into the political performance appraisal system to balance local governments' tendency between ecological protection and economic growth [190], increase the percentage of environmental protection and pollution control in the performance appraisal, and take environmental quality as an important indicator of official promotion assessment. Only the establishment of such a long-term government can guarantee the functioning of the mechanism for government-led voluntary supply by residents.

10.3.2 Design of a Rational "Intergenerational Contract", with Consideration Given to Intergenerational Equity

Spatial and temporal externalities need to be considered in the supply of environmental public goods, thus it is necessary to design proper environmental protection plans for both enterprises causing pollution and residents with a WTP for environmental protection. In terms of enterprises, the government should not only forbid them to discharge pollutants at will, but also adopt measures to encourage them to research and develop new technologies; as regards residents, even if present generations are willing to pay for environmental protection, the government should give compensation to their offspring so as to effectively motivate more residents to voluntarily contribute to environmental protection, and solve the "free-rider" problem. Zeng and Xia [191] proposed that the collaborative innovation on environmental technology based on government leadership could effectively reduce the transaction cost of environmental technology R & D; the coerciveness based on environmental regulation could effectively reduce the execution cost of environmental technical services; the public participation based on community residents could effectively reduce the supervision cost of such services.

1. Use of the policy mix to "internalize" the "spatial" and "intergenerational externalities" of polluting enterprises

On the one hand, Pigovian taxes can be imposed on polluters failing to meet the national standards for discharge of "three wastes", and tax revenue can be used as part of the "environmental protection fund" for pollution control. On January 1, 2018, China began to implement the *Regulation on the Implementation of the Environmental Protection Tax Law of the People's Republic of China*, which meant that the system of fee charging for pollutant discharge lasting 38 years withdrew from the stage of history. Enterprises and public institutions both should pay "environmental protection taxes" for discharging air, water and solid waste, or causing noise.

On the other hand, for enterprises which have applied new technologies and greatly improved environmental quality, the government should give rewards. Rewards and punishments should be clear. For enterprises which have made technological progress, the government should give subsidies to motivate them to make improvements. The policy mix combining punishments and subsidies is beneficial to the upgrading of industrial structure, and the survival of the fittest in the business circles, and the acceleration of economic growth. This is consistent with the research conclusion drawn by Li Dongdong et al. [118]. According to the research conclusion drawn by Li and Niu [192], technological progress can significantly reduce the CO_2 emission in the current period, and plays a significant role in energy conservation and emission reduction. In the view of Chen Shiyi, it is advisable to energetically develop clean coal technology and clean energy, and encourage the development of new energy and renewable energy sources so as to adjust the energy consumption structure and reduce carbon emissions. In the view of Zeng and Wang [194], the environmental regulation based on market incentives has a better effect on motivating enterprises to absorb environmental technology, compared to that based on control commands.

As regards concrete policies, direct monetary subsidy is not the only way to stimulate enterprises, and brainstorming may be employed. For example, for enterprises with technological progress, the government may provide some preferential policies, shorten their loan approval period, or directly provide a green channel that allows them to have priority to apply for loans.

2. Provision of "intergenerational subsidies" to residents so as to promote intergenerational equity

The survey results suggest that over half of residents are willing to pay the environmental protection fund for the sake of their own health and offspring, but the government should give priority to and solve problems on how to motivate more residents and even the entire people to participate in pollution control, and how to ensure that this fund could be put into proper use, which is a concern of residents paying the fund.

As regards strategies, the "intergenerational contract" idea presented by Kotlikoff et al. [185] may be used for reference. If the environmental protection fund paid by the generation t (young and middle-age period) could enable the generation $t + 1$ (old-age period) to obtain compensation, the above problems would not exist. For residents paying the environmental protection fund, we suggest giving subsidies (intergenerational transfer) to them in their old-age period, and the specific subsidies may be pension subsidies or medical insurance subsidies, or inheritance tax credit policies (if any). In short, the principle of intergenerational subsidies should be based on the hypothesis of economic man to enable residents to maximize utility through lifetime consumption and environmental quality, give discount compensation to those residents volunteering to pay for environmental protection in middle age when they reach old age, and allow their benefits to reach their opportunity costs at least.

Radner [175] has proved that if the discount factor is big enough to make the principal and the agent have sufficient confidence, the Pareto first-order optimal risk sharing and incentive can be achieved. This also means that the government's fulfillment of commitments is particularly important. Although a considerable number of residents who voluntarily paid for environmental pollution control when they were young, but according to the hypothesis of economic man, it is impossible to meet the huge capital demand for pollution control in reliance upon residents' willingness alone. Even if the "altruistic motivation" is taken into account, Becker [35] pointed out that altruism means that the outcome of individual behavior can bring benefits to others, but in fact, the motivation thereof is still self-interested. Thus, when developing policies, the government should fully show its leading role, and clearly reveal residents' costs and future benefits by the "intergenerational contract" to enhance their confidence.

10.3.3 Introduction of a Competitive "Environmental Protection Fund Company" to Avoid Government Failure

To ensure the effective functioning of the mechanism for government-led voluntary supply of environmental public goods by residents, we must establish a clean and efficient service-oriented government, and use the incomplete contract theory to combine the interests of all parties in society. There is an important question worth considering. Who shall manage the "environmental protection fee" paid by residents and enterprises, and who shall have the disposal right? If the fee is directly delivered to the government sector, the previously discussed "government failure" problem will be inevitable.

As regards this issue, this book suggests setting up a third-party company—"Environmental Protection Fund Company", which should be selected by competitive means, establish principal-agent relations with the government, manage the environmental protection fund under the entrustment of the government, and take charge of the fund operation and the preparation and implementation of environmental governance plans. The government, residents and enterprises all can supervise it. The fund mainly comes from the environmental protection fee voluntarily paid by residents, and the pollution tax paid by enterprises. It is used to curb environmental pollution, is a source of "subsidies" for residents in period $t + 1$, and used as an incentive for enterprises making technological improvement to the benefit of sustainable development.

Here we consider establishing an "Environmental Protection Fund Company" by market means, with a hope that the company could run efficiently to avoid government failure. The operation of the company is market-oriented. The subjects of society—residents and enterprises are supervisors; as the principal of the fund, the government is also the owner thereof, and should be responsible for the ultimate costs and benefits.

10.3.4 Enhancement of Environmental Protection Education, Popularization of Environmental Protection Knowledge

Hou [195] indicated that public participation in the environment included many aspects: first, participation in the management and decision-making of the government's environmental affairs, which meant that the public should have the basic right of knowing environmental information, the right to participate and the right of action; second, citizens' attitude towards and behavior of participation in environmental protection or low-carbon practice.

The above discussion has suggested that if residents are aware of the seriousness of environmental pollution, maintain good personal environmental awareness, have the enthusiasm for participating in environmental protection activities, know more about environmental protection knowledge… then our environmental protection will become much easier, and the supply of environmental public goods will be more effective.

In 2010, CGSS specifically put forward a question concerning environmental protection: "Which of the following ways do you think is the best way for Chinese public and their families to protect the environment?" Unexpectedly, only 24.27% of respondents chose "severely punishing enterprises causing environmental damages"; 38.37% chose "providing individuals more information on the benefits of environmental protection"; 15.29% answered "cannot choose". As regards the survey of another question, 31.29% of respondents considered "I could hardly figure out whether my current lifestyle is good or bad for the environment", and only 4.64% considered that they knew exactly whether their lifestyle is bad or good for the environment. These survey results show that more residents hope to obtain more information on environmental protection.

Chinese residents' environmental awareness and knowledge level are generally low; the overall level of public participation in environmental protection activities is low, and the participation mode is single. According to survey findings, only when environmental pollution directly damages personal interests can more people are willing to take action. In other words, there is only terminal participation, but a lack of planned participation and process participation, with weak individual behavior participation. Most of the public believe that environmental protection is just a task of the government. Therefore, China should energetically develop environmental education, especially social environmental education, and add the content thereof into news media to enrich environmental information. Besides, the mechanism for public participation in environmental protection through the "environmental impact assessment system" remains to be implemented. The Japanese public utilizes environmental dispute litigation and political votes to urge enterprises and the government to curb pollution, which provides some references for China.

It is reported that in 2015 China's environmental protection departments at all levels completed a total of 2345 environmental education bases, organized and carried out social environmental publicity and education activities for 12,175 times, with more than 30.345 million participants in such activities. The government may make the utmost of We-Media platforms to provide public opinion guidance and science popularization, e.g. using WeChat, Weibo and other We-Media platforms to organize science popularization education on environmental protection knowledge, or to guide residents' in the correct direction of public opinion. We are strengthening the popularization of environmental protection knowledge. Only when the people enhance their environmental awareness, and improve their environmental knowledge level can they turn their environmental protection thought into action.

10.3.5 Improvement of Public Participation and Information Disclosure Systems

At present, China's public participation systems in connection with pollution prevention and control of the manufacturing industry include letters and visits (public complaints and proposals), reports and complaints, as well as hearings. Since 1993, China has issued a series of environmental assessment and letters and visits policies to encourage the public to participate in pollution control. For instance, the *Notice by the National Environmental Protection Agency, the State Planning Commission, the Ministry of Finance, and the People's Bank of China of Strengthening the Administration of the Environmental Impact Assessment on Construction Projects with Loans Granted by International Financial Organizations* issued in June 1993 indicates that the modes of public participation in environmental impact assessment mainly include hearings; the *Measures for the Administration of the Expert Bank for Environmental Impact Assessment Review* issued in September 2003 determines a random mechanism for expert selection, and develops the dynamic management measures for the expert bank; the *Interim Measures for Public Participation in Environmental Impact Assessment* issued in 2006 defines the specific scope, procedures, mode and period of public participation in environmental impact assessment. With regard to letters and visits, the *Notice by the General Office of the CPC Central Committee and the General Office of the State Council on Further Strengthening Public Complaints and Proposals* issued in 1989 proposes improving the system of environmental public complaints and proposals, and strengthening the information feedback and the supervision and handling of cases with regard to public complaints and proposals.

Public participation is not only a basic principle, but also a basic system in the environmental field of developed countries. Overseas scholars introduced "public participation" from the political field to social management, not only carried out a lot of outstanding researches on basic theories, e.g. the democratic theory, the public choice theory, the public trust theory, and the theory of environmental rights, but also launched the full practice for public participation in environmental governance earlier. Many developed countries, e.g. the US and Canada, have determined an all-round and complete legal system for public participation based on environmental rights for the purpose of implementing the public participation principle.

In China, public participation has not yet been determined as a basic principle in the field of environmental protection, and public participation in environmental protection lacks basic institutional guarantee. However, as environmental problems have become a growing concern for the government and the society, the requirements for public participation in environmental protection have been elevated, and the masses' participation in environmental protection by effective and legal means has been raised to the height of upholding social equity and justice, and building a harmonious society.

China's information disclosure system in relation to pollutant discharge from the manufacturing industry was established relatively later, and mainly targeted at listed companies. In 2005, the State Council issued the *Decision of the State Council on Implementing the Scientific Outlook on Development and Strengthening Environmental Protection* (hereinafter referred to as the *Decision*), indicating that enterprises should make environmental information public, and receive public views with regard to the development planning and construction projects of public environmental rights and interests in the forms of hearings, demonstration meetings or public notices. It should be noted that the *Decision* did not compulsively require enterprises to disclose environmental information. In 2007, the State Environmental Protection Administration (present-day Ministry of Ecology and Environment) developed the *Measures for the Disclosure of Environmental Information (for Trial Information)*, and defined the content of information disclosure from the aspects of the government and enterprises. The disclosure of governmental environmental information covers the scope, procedures and mode of disclosure; the disclosure of corporate environmental information includes the content of environmental information and the time of disclosure. The government would give rewards to those enterprises volunteering to disclose environmental information. In 2008, the *State Environmental Protection Administration* issued the *Guiding Opinions on Strengthening the Environmental Protection Supervision and Management of Listed Companies*, encouraged listed companies to routinely and voluntarily disclose other environmental information, and spurred enterprises to take the initiative to assume social environmental responsibility.

We may learn from the information disclosure systems of some developed countries.

First, the US. In 1969, the *National Environmental Policy Act* (NEPA) enacted by the US stipulated that environmental impact statements and the opinions of relevant authorities shall be made public pursuant to the *Freedom of Information Act*. The *Regulations for Implementing the Procedural Provisions of NEPA* enacted later specified information disclosure. In the 1980s, the US enacted the *Emergency Planning and Community Right-to-Know Act* and the *Toxic Chemical Release Inventory*, published the national "Toxic Release Inventory Fact Sheet", and disclosed the emission and disposal information of more than 300 hazardous substances which enter air, water and landfills, and are transported to off-site places. In 1996, the US enacted the

Electronic Freedom of Information Act, providing that the government shall disclose environmental protection and other administrative information.

Second, Japan. Its environmental information disclosure is focused on technology and data, environmental policies, as well as environmental behavior and effect. As regards the disclosure of environmental technical information and environmental policies, the Japanese government will provide technical assistance, technical exchanges and cooperation in air pollution prevention and control, water quality improvement, and pollutant disposal, and partially disclosed technical information; disclose environmental technologies and policies via the Internet; allow the public to keep abreast of environmental policies, as well as the environmental behavior and effect of the central and local governments and enterprises through announcements.

Third, Germany. Germany's disclosure of environmental pollution information is manifested in its *Environmental Information Act*. The Act clearly defines the concept of environmental information, and specifies the public right of knowing environmental information, the government's duty to disclose environmental information, the content of environmental information disclosure, the procedures and contents for the public application for obtaining environmental information, etc. The information disclosure covers waters, air, and soil.

Only when public participation systems and the information disclosure system are improved can China's supply of public goods be effectively achieved.

Chapter 11
Path for Chinese Residents' Participation in Supply of Environmental Public Goods

There are many modes for residents' participation in environmental governance, including participating as individuals or families, voicing opinions via media, or participating in the form of environmental NGOs. The participants in environmental governance in various capacities refer to the basic units capable of participating in and acting on social development, including natural persons, legal persons (e.g. social organizations, Party and government organs and public institutions, NGOs, Party and mass organizations, non-profit organizations, and enterprises) and the media, which are collectively known as residents herein.

Environmental governance is a systematic project. Residents play a significant "coordination" role in environmental governance, and are an important constructive force for improving the performance and level thereof.

11.1 Significance of Residents as the Environmental Supply Side

Due to the game between the government and enterprises, system obstacles and other realities, the effect of a complete set of environmental governance measures based on command-control is insignificant. In recent years, lots of facts have suggested that residents' participation has been playing an increasingly important role in environmental governance.

This is closely related to the impact of environmental pollution on residents' life, which awakens their environmental awareness. But residents' behavior and collective actions arising from the awakening may have complicated effects in both positive and negative directions, and the underlying motivations thereof are noteworthy.

J. Yang, *Effective Supply Mechanism and Path of Environmental Public Goods in China*,
Public Economy and Urban Governance in China,
https://doi.org/10.1007/978-981-99-2939-9_11

Environmental pollution has led to a huge change in residents' environmental awareness. More and more residents have begun to doubt the reliability of government provision of public goods, and even the rationality of the government serving as the supplier of public goods. The traditional consensus has been challenged. In other words, the clean environment has been no longer a by-product of economic growth, but an important foundation for guaranteeing the people's prosperous and contented life, together with technological or literary creation. Residents' participation in environmental governance has a good driving effect, but if residents participate in environmental governance by an improper approach, social problems will arise. An equilibrium mechanism for supervision and restriction composed of "the government, social organizations, residents, and the media" appears particularly important to prevent mass incidents. The real cure is building a mechanism of government-enterprises environmental responsibility coordination. Besides, it is undoubtedly important and urgent to mobilize and motivate residents to advance the responsibility coordination, and to build a corresponding mechanism.

Enabling residents to jointly provide environmental public goods, and be the deciders of the quantity of public goods so as to form cooperative governance is also a way to effectively solve environmental pollution. Scholars stressed that environmental protection was focused on taking residents' willingness into account. Residents' cognitive change plays a positive role in promoting environmental governance. Only when the channel of residents' participation in green development is expanded, and their cognitive state and the cost of environmental governance are taken into full account can effective governance be achieved. Actively building a multi-subject environmental collaborative governance system with the joint participation of residents and social organizations is an inevitable choice to ensure the improvement of environmental quality.

If residents participate in environmental governance by an improper approach, social problems will arise. The "Collective Walking" incident in Xiamen in 2007, the Dalian PX (paraxylene) incident in 2011, the mass protest against Hongda molybdenum-copper project in Shifang, Sichuan after 2012, the mass protest against seaside sewage discharge in Qidong, Jiangsu, the mass protest against PetroChina's PX refining project in Kunming, Yunnan, as well as other wide-reaching environmental mass incidents marked that China entered an "environment-sensitive period". Feng Jie et al. held that environmental mass incidents, illegal land expropriations and housing demolitions, as well as labor disputes have become the "Three Carriages" causing mass incidents.

As regards the frequent occurrences of environmental mass incidents, the reasons behind them should be intensively researched. First, China lacks a positive and effective mechanism for residents' participation in the environment; second, from the perspective of residents, the government and enterprises are an alliance, and only mass incidents could force polluters and the government to satisfy residents' environmental appeals; third, the government officials' notion of "Growing First, Cleaning Up Later" and the imperfect environmental legal system have caused some problems. It follows that an equilibrium mechanism for supervision and restriction composed of "the government, social organizations, residents, and the media" appears particularly important to prevent mass incidents.

11.2 Effect and Obstacle of Residents' Participation in Supply of Environmental Public Goods

As the old saying goes, "When everybody adds fuel, the flames rise high," if the entire people jointly participate in environmental governance, all difficulties will be solved easily.

11.2.1 Effect of Residents' Participation in Supply of Environmental Public Goods

1. Promoting the social co-governance mechanism of government-enterprises environmental responsibility coordination

How will we break through the environmental management modes concerning the "government regulation", the "market mechanism" and the "voluntary organization self-governance", eliminate the government-based thought that "only the government can execute management, and perform well in management", and realize the mechanism innovation research on environmental governance?

Residents' participation in environmental governance is conducive to inspecting multiple realistic approaches and schemes for residents' participation, government management, as well as government-enterprise governance, analyzing respective advantages and disadvantages, and helps us to explore an effective form for encouraging the participation of individual residents, associations, societies, and other social forces and voluntary organizations, especially to explore how to motivate social forces to participate in the construction of environmental public goods by system design. In addition, residents' participation is conducive to thoroughly probing into the respective boundaries and specific connecting patterns of rigid (administrative approval, licensing, punishment, etc.) and flexible environmental governance (administrative guidance, advice and suggestions, contracts, rewards, subsidization, etc.). Rigid environmental governance is mainly implemented by the government by use of public power, including curbing loans, with the compelling force of law as the guarantee. Flexible environmental governance includes strengthening green education or popularizing science and technology to improve residents' environmental awareness, with a focus on guiding residents from "passive" to "active environmental protection", and social organizations from "negative opposition" to "positive collaboration". It is advisable to explore the specific feasible approaches of setting up a "multi-sector (cross-sector) cooperation" mechanism or "integrated environmental service management" agency; with environmental governance aimed at quality environment, carry out inter-departmental work concerning environmental protection, industry and commerce, tax and education departments to overcome the drawbacks concerning separate departments and decentralized authority, and provide reference schemes for the government to truly attach equal importance to and combine services and governance.

2. Promoting the social supervision mechanism of government-enterprises environmental responsibility coordination

To achieve the win–win results of economic growth and environmental protection and build ecological civilization and "Beautiful China", we need not only the theoretical guidance—"lucid waters and lush mountains are invaluable assets", but also residents' participation spirit—"I'm in charge of my environment" as practical support.

As a subject of environmental protection, residents must participate in environmental governance. Everyone is a participant in environmental governance, but due to the imperfect mechanism for residents' participation in environmental governance, there is no effective social supervision mechanism. At present, the environmental information disclosure of the Chinese government is incomprehensive, untimely and inaccurate, and most polluting enterprises are unwilling to disclose their data. To achieve greater results, we should effectively mobilize social forces to participate, break the original pattern of "active government, passive enterprises and inactive residents" in China's environmental governance, and motivate residents, social organizations, the media, environmental NGOs, and so forth to make concerted efforts to form mutual restrictions and supervision with the government and enterprises.

11.2.2 Main Obstacles of Residents' Participation in Supply of Environmental Public Goods

1. System obstacle

The system factor is the key factor influencing the performance and success or failure of environmental governance. The formation and effective functioning of a government-led innovative system and mechanism with multiple participants and effective interactions needs not only the changes in thinking, value orientation and behavior and standpoints of the government sector and related interest groups, but also a sound and secure system.

In terms of the government, it is faced with how to transform from "focusing more on fairness" under the traditional governance mode to "balancing fairness and efficiency", from "process orientation" to "result and process orientation", and from "administrative orientation" to "market orientation"; on the institutional level, it should achieve the structural transformation from a single authoritative subject to multiple subjects; most importantly, on the operational level, it should change from the means dominated by control and command and supplemented by economic means to such means as equal communication, exhortation and voluntariness. It is necessary to establish the notion that "management is service", and to implement the "people-oriented" governance concept and the Scientific Outlook on Development.

In terms of enterprises and factories, they are faced with how to cultivate a sense of social responsibility and perform the basic obligations for environmental quality, and how to transform the environmental governance idea from "being dominated by control" to "being dominated by prevention". As regards residents, their participation willingness, participation expression and participation right for environmental governance are limited, and their environmental awareness is still in the development stage, and has not become an internalized awareness. At present, residents are faced with how to actively participate in environmental governance. As for social organizations, they are faced with the establishment of legal thoughts and rule awareness, and the improvement of overall quality. In this aspect, researches should be focused on exploring the implementation approaches and modes for communication and cooperation, consultations and dialogues, as well as partnerships among related social interest groups, analyzing the premises, basis and specific conditions for different groups to form consensus and common goals, and enabling the government, enterprises, residents and social organizations to establish open, equal, inter-conditioned, harmonious and interactive relations.

2. Residents' environmental awareness obstacle

CGSS conducted a survey on residents' environmental protection activities, with the content covering residents' environmental protection behavior: e.g. waste sorting, BYO eco-friendly shopping bags, and reuse of plastic bags; environmental awareness: e.g. discussing environmental protection with relatives and friends, taking an active part in environmental publicity and education activities organized by the government and units, taking the initiative to pay attention to environmental issues reported by the media, and actively participating in complaints and appeals with the request for solving environmental problems; as well as the subjective initiative of environmental protection behavior: e.g. making donations to environmental protection, and planting trees at their own expense. See Table 11.1 for details. In the table, the sample size is 11,438 persons. With respect to the answers to 10 questions, respondents answering "often" to the question—"Reuse plastic bags" accounted for the highest percentage, 49.25%; followed by those answering "often" to the question—"Bring your own shopping basket or bag when shopping for daily necessities", accounting for 39.46%; but the respondents answering "often" to the remaining 8 questions only accounted for a small percentage, and most answered "never or occasionally". It shows that for residents' participation in environmental protection, their environmental awareness should be the focus of attention.

Table 11.1 Statistics on residents' environmental protection activities

SN	Activity or behavior	Often		Never or occasionally		Other (Refuse or do not know)
		Number of respondents	Percentage %	Number of respondents	Percentage %	Percentage %
1	Waste sorting	1404	12.27	10,012	87.53	0.19
2	Discuss environmental protection with relatives and friends	876	7.66	10,534	92.10	0.24
3	Bring your own shopping basket or bag when shopping for daily necessities	4514	39.46	6899	60.32	0.22
4	Reuse plastic bags	5633	49.25	5776	50.50	0.25
5	Make donations to environmental protection	225	1.97	11,181	97.75	0.28
6	Take the initiative to pay attention to environmental issues reported on radio, TV, and in the press	1459	12.76	9949	86.98	0.26
7	Take an active part in environmental publicity and education activities organized by the government and units	452	3.95	10,953	95.76	0.29
8	Take an active part in environmental protection activities held by nongovernmental environmental groups	270	2.36	11,133	97.33	0.31
9	Maintain trees or green space at your own expense	436	3.81	10,982	96.01	0.17

(continued)

Table 11.1 (continued)

SN	Activity or behavior	Often		Never or occasionally		Other (Refuse or do not know)
		Number of respondents	Percentage %	Number of respondents	Percentage %	Percentage %
10	Actively participate in complaints and appeals with the request for solving environmental problems	174	1.52	11,237	98.24	0.24
Total number of respondents 11,438						

Data source Summarized according to the answers to the B22 survey questions of CGSS 2013

11.3 Strategy Study for Residents' Participation in Supply of Environmental Public Goods

11.3.1 Issues to Be Reviewed for Residents' Participation in Environmental Governance

"Lucid waters and lush mountains are invaluable assets." China has been fully fighting "three defense battles concerning blue skies, clear waters, and clean lands". To ensure victory, it is of great importance to establish a mechanism of government-enterprises environmental responsibility coordination. What are the social forces participating in environmental governance? What are the roles and functions of social organizations, residents (including experts) and the media in environmental governance? How do we motivate social organizations and residents to jointly participate in the environmental governance system? How do we enable social forces to play a significant role in propelling the building of the mechanism of government-enterprises environmental responsibility coordination? This is the objective of this study. Focusing on this objective, this part will intensively clarify the following specific issues.

1. What are the social forces participating in environmental governance? What are their roles and functions?

At present, the participants in social environmental governance mainly include the government and enterprises. In the traditional environmental governance mode, the government is generally the single center and subject, and other social subjects are supervisees and order followers. According to Samuelson's theory, the environmental goods including clean air, clean water and clean soil provided by environmental governance belong to a kind of pure public goods. Due to the public good attribute of environment, the "free-riding" behavior is inevitable. What are the social forces participating in environmental governance? What are the roles and functions of social organizations, residents (including experts) and the media in environmental governance? These are realistic issues to be clarified first in this study.

2. Mode selection for the participation of social forces including residents in the mechanism of government-enterprises environmental responsibility coordination

In the evolution process of environmental governance, there are three typical governance modes: first, the government regulation mode with Pigovian taxes as the institutional arrangement; second, the market governance mode based on the institutional arrangement of the Coase theorem; third, the voluntary governance mode with the environmental civil society theory as the institutional arrangement. However, practice suggests that due to the complexity of environmental problems, government regulation failure, market regulation failure and social voluntary failure have successively occurred in environmental governance. As environmental quality is the most urgent concern for such subjects as residents, social groups, voluntary organizations and other social forces, their participation mode of environmental governance needs to be rebuilt, and social environmental governance needs a new public governance mode. How will we build a social co-governance mode? How will social forces participate? These are core issues to be solved in this study.

3. What channel should social forces build so as to achieve the effective supervision on the environmental protection behavior of the government and enterprises?

In the process of participating in environmental governance, in what ways should social organizations, residents and the media disclose environmental protection information in relation to the government and enterprises? How will the accuracy, timeliness and effectiveness of such information be ensured? What are the roles of social forces? How will these subjects promote the cooperative governance of the government and relevant enterprises in the supervision process?

4. Implementation path and strategy for the environmental responsibility coordination mechanism secured by social forces

How will we motivate residents, social groups, voluntary organizations and other social forces to participate in environmental governance, to serve as the subjects of environmental responsibility coordination together with the government and enterprises, and to become the core factors of the coordination mechanism? How will we design the mechanism factor to enhance residents' awareness of responsibility in environmental governance coordination, and cultivate their sense of ownership? These are the strategies to be solved herein.

11.3.2 Path Exploration for Residents' Participation in Supply of Environmental Public Goods

1. Main mode for promoting the supply of environmental public goods

This study is focused on how to break through the traditional environmental management modes concerning the "government regulation", the "market mechanism" and the "voluntary organization self-governance", eliminate the government-based thought that "only the government can execute management, and perform well in management", realize the government's mechanism innovation on environmental governance, and build a "cooperative co-governance mode" for government-enterprises environmental responsibility coordination with the participation of social organizations, residents and the media.

As a comprehensive governance mode for social public affairs, the cooperative co-governance mode shows a multi-subject trend. In addition to the government and public institutions, NGOs, for-profit/non-profit organizations, civil organizations, private groups, residents, and so forth all can participate in public governance as the co-subject thereof, i.e. "cooperative co-governance". It has the following basic characteristics: first, partnership with multi-subject participation; second, attaching importance to result and resident orientation; third, attaching importance to contract-market relations; fourth, structurally showing a flat network governance system; fifth, emphasizing democratization and valuing residents' participation. Thus, public governance and "cooperative co-governance" on this basis provide significant enlightenment for the participation of social forces in environmental governance.

(1) Exploration of a path for residents' shift from passive to active environmental protection

It is advisable to intensify green education or popularize science knowledge on environmental protection by educational publicity so as to improve residents' environmental awareness, and guide them from "passive" to "active environmental protection". We suggest schools offer courses related to environmental protection since kindergarten to truly enable education to go first, and allow children to have environmental awareness and action from an early age. Besides, we suggest giving play to the publicity role of community neighborhood committees, and implementing environmental protection points systems for residents in different regions. For example, in terms of waste sorting and BYO eco-friendly shopping bags when shopping in vegetable markets, residents who have such green behaviors will be rewarded with environmental protection points, which can be used to exchange for basic necessities, thereby motivating the vast majority of elderly people to voluntarily join the environmental protection team. As for the middle-aged people, we encourage linking up their environmental behavior with credit score (positive or negative), which may be associated with individual credit approval, exit visa, children's choice of school, or even used to offset some demerit points for motor vehicles.

(2) Exploration of a participation mechanism of social organizations from disorderly to orderly

Social organizations play a significant role in environmental protection publicity and education, environmental policy participation, environmental behavior supervision, and other aspects. On the one hand, it is advisable to actively promote the development of environmental protection organizations, loosen the management on administrative examination and approval thereof, and enhance financial support for them to provide broader room for their development. On the other hand, to avoid environmental mass incidents, and deal with various contradictions and conflicts of interest, we suggest building a local multi-subject environmental consultation and dialogue platform, employing the joint conference mode to enable environmental protection organizations to fully communicate information and express interest demands so as to work out a unified plan. This can ensure the orderly and civilized participation of environmental protection organizations in environmental governance.

(3) Exploration of system design for the media to release reliable information

As special organizations, the media have unique advantages and effect. On the one hand, it is advisable to continue to expand the freedom right of the press, promote the supervision, guidance and communication power of the press and media, and strengthen the intermediary role thereof in information spreading and disclosure; on the other hand, it is necessary to establish and journalism norms, implement effective punishment and governance for such phenomena as media hype and false publicity, and guide the press and media to provide accurate, effective and timely information in the governance process.

2. Supervision mechanism for residents' participation in environmental public goods

In the past, China's ecological environment supervision had been dominated by government regulation. To improve the self-restricting mechanism for polluters and achieve a qualitative leap from "being required to abide by laws" to "take the initiative to abide by laws" in the entire society, social supervision must be intensified. Promoting and improving the social supervision mechanism is an important part of residents' participation in environmental governance.

(1) Establishment of a channel for corporate environmental information disclosure

It is advisable to establish a system for encouraging residents, the media, environmental NGOs and other social organizations to urge the disclosure of corporate environmental information; strengthen the social supervision role of residents and social organizations in the full life circle of enterprises concerning project-based environmental impact assessment, project construction, terminal pollution discharge and governance, and spur enterprises to conscientiously fulfill their environmental responsibility.

(2) Improvement of a mechanism for expanding the subject scope of environmental information disclosure

It is advisable to introduce social forces to expand the subject scope of environmental information disclosure, change the status that only the environmental protection authorities at or above provincial level are allowed to disclose environmental information in the form of communiqués on a regular basis to ensure the integrity of environmental information, enable all enterprises to be under supervision, and achieve information symmetry. Besides, in the process of environmental assessment on enterprise projects, social forces may intervene in supervision by urging the government to hold hearings, symposiums and expert workshops.

(3) Method of setting up an environmental information sharing platform

It is advisable to give full play to the role of social forces including the media to supervise the government sector in building an environmental network information platform, and establishing an interregional information notification system for timely communication on latest pollution and environmental information and enhancement of interregional cooperation, thereby boosting the real-time information sharing nationwide.

3. Implementation path for residents to promote the effective supply of environmental public goods

Under the pressure-type system of the Chinese unitary government, there is limited space for institutional innovation. With the help of the previously identified channels for residents' participation in environmental responsibility coordination and the mechanism obstacles, targeted policy suggestions are put forward. The author intends to carry out a differentiated policy research from the following perspectives.

(1) Government-enterprise-society "tripartite linkage" in environmental governance, with accurate orientation and clear relations

It is necessary to fully understand and determine the differentiated responsibilities and functions of the government, enterprises and society in environmental governance, and attach importance to the difference between regions and space. The key is establishing a government-enterprise-society linkage mechanism, and forming a "combined power" for participation in environmental governance by integration on this basis. As regards the paths of residents' participation, we must pool the power of the government, science, systems and the masses, further advance theoretical, practical and institutional innovation, thoroughly break through the collective action dilemma, vigorously build a governance system of government-enterprises environmental responsibility coordination with the joint participation of residents and social organizations, and form the most extensive united front for environmental protection so as to accelerate the formation of a co-governance pattern where the entire people "breathe the same air and get it done together".

(2) Cooperative co-governance, residents' organized participation path

First, public participation from "passive" to "active". The disorderliness of individual residents' participation can be overcome by organized participation. The main paths include: arranging an information supply system in residents' participation process so as to consolidate the basis of participation, and establishing a government response and judicial relief mechanism with the participation of residents so as to elevate the effectiveness of participation.

Second, participation of social organizations from "disorderly" to "orderly". This includes: (a) fully cultivating and incubating social organizations and various NGOs related to environmental protection to expand the voicing subject scope of social organizations; (b) setting up all sorts of platforms and unimpeded channels to allow social organizations to fully participate in all processes and links of environmental governance, and to achieve whole-process participation; (c) continuing to strengthen the cooperation among social organizations, the government and enterprises, to realize the organizational function of residents in the atomized state, and to form positive constructive forces for environmental governance.

Third, media voicing from "criticism" to "construction". Full play should be given to the positive constructive force of all sorts of emerging media including websites, WeChat and client side (in the We-media era) in environmental governance. This requires not only active voicing, i.e. incorporating the media force into environmental governance, but also active construction, i.e. influencing and shaping the citizens' environmental awareness and the credibility of social organizations, publicizing government policies, creating a good public opinion atmosphere for environmental governance, and providing public opinion support for "tripartite" participation therein and resultant force.

(3) Classified policy implementation, development of differentiate policies for resident's participation

This requires dividing the country into different regions (cities), conducting a systematic survey according to the economic development degree and environmental governance pressure of different regions (cities), summarizing corresponding prominent problems and typical characteristics on this basis, and adopting different solutions and mechanism designs based on the principles of putting people first, adjusting measures to local conditions, etc. according to the education level, maturity of citizens, cognitive level of residents, and social forces' willingness and ability to participate in environmental governance in different regions (cities).

11.3.3 Problems of Residents' Participation in Supply of Environmental Public Goods

1. Inconsistent environmental notions and awareness of residents nationwide

Due to the differences in economic income, economic development modes and educational attainments, residents have different attitudes and behaviors with regard to environmental protection. In short, higher-income and higher-educated residents show more outstanding enthusiasm and practical action for environmental protection. In Table 11.2, a questionnaire survey on residents' WTP for environmental protection is conducted. The statistical results suggest: for the question "To what extent are you willing to pay higher prices for environmental protection?", the percentage of respondents nationwide choosing "strongly willing" and "relatively willing" was 42.32%, while the percentage in Shanghai was 48.4%, 6% higher than the national average; for the question "To what extent are you willing to pay higher taxes for environmental protection?", the percentage of respondents nationwide choosing "strongly willing" and "relatively willing" was 32.93%, while the percentage in Shanghai was 34.4%, nearly 2% higher than the national average.

The above survey statistics suggest that for cities like Shanghai, due to their relatively faster economic development, the overall educational attainment of residents therein is higher than the national average, and they are more concerned about environmental protection, with advanced environmental behavior and awareness.

Figure 11.1 comparatively displays the residents' WTP for environmental protection in three first-tier cities—Beijing, Shanghai and Guangzhou. Compared to the national average, residents in the three cities are willing to pay higher prices for environmental improvement. Data analysis suggests that the residents of Guangzhou have the strongest WTP; the residents in Beijing and Shanghai have a similar willingness, both far higher than the national average.

Table 11.2 Statistics on residents' WTP for environmental improvement

Question description	To what extent are you willing to pay higher prices for environmental protection?		To what extent are you willing to pay higher taxes for environmental protection?	
Option	National average	Shanghai	National average	Shanghai
Strongly willing	8.50	7.64	4.58	5.10
Relatively willing	33.82	40.76	28.35	29.30
Neither willing nor unwilling	18.33	19.75	18.85	21.02
Less willing	23.01	10.83	27.89	22.29
Strongly unwilling	7.05	9.55	8.71	10.19
Unable to choose	8.85	11.46	10.19	12.10
Sample size	–	–	3672	157

Data source Calculated according to the data of CGSS2010
Unit: %

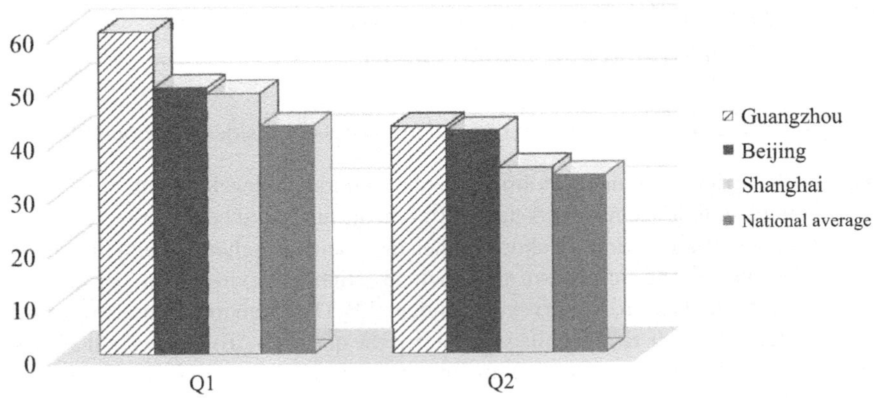

Q1: To what extent are you willing to pay higher prices for environmental protection?

Q2: To what extent are you willing to pay higher taxes for environmental protection?

Fig. 11.1 Comparison of residents' WTP for environmental protection in Beijing, Shanghai, Guangzhou and other parts of China

Table 11.3 lists four questions in connection with environmental protection, and mainly surveys residents' environmental awareness in normal life. Table 11.4 lists the survey results of the four questions. For Question 1: "Do you often sort glass, aluminum cans, plastic or newspaper, etc. for ease of recycling?", the percentage of residents in Shanghai answering "Always" was 25.48% while the national average was 11.93%; the percentage of residents in Shanghai answering "Often" and "Sometimes" was also far higher than the national average. The statistical results of the answers to the remaining questions were consistent with that of Question 1. It could be inferred that the environmental awareness of residents in Shanghai was significantly higher than the national average.

Figure 11.2 comparatively researches the environmental awareness of residents in three first-tier cities—Beijing, Shanghai and Guangzhou, and other cities of China. The data therein displays the percentage of respondents answering "Always + Often" to the four questions in Table 11.3. The statistical data suggests that the environmental awareness of residents in any of the three cities was higher than the national average.

Table 11.3 Questions on environmental awareness of residents in Shanghai

Question No	Question description
1	Do you often sort glass, aluminum cans, plastic or newspaper, etc. for ease of recycling?
2	Do you often reduce the household consumption of oil, gas, electricity and other energy sources or fuels for environmental protection?
3	Do you often save or reuse water for environmental protection?
4	Do you often choose not to buy any products for environmental protection?

Table 11.4 Statistical analysis on residents' environmental awareness

Question and option		Always	Often	Sometimes	Never	No recycling system in the domicile nearby
Q1	National average	11.93	19.72	23.75	17.35	25.88
	Shanghai	25.48	18.47	22.93	14.65	
Q2	National average	9.86	22.41	39.79	25.77	
	Shanghai	21.66	24.20	35.67	18.47	
Q3	National average	17.21	31.54	33.82	15.94	
	Shanghai	31.85	29.94	26.11	12.10	
Q4	National average	7.30	15.69	41.07	33.61	
	Shanghai	16.56	21.66	38.22	21.66	

Data source Calculated according to the data of CGSS2010
Unit: %

Especially for residents in Beijing, they ranked first in terms of answering "Always + Often" to the four questions. This indicated that the residents in Beijing had strong environmental awareness and environmental action.

Table 11.5 compares the environmental behavior of residents in Shanghai with the national average. The data suggested that the actual environmental behavior of residents in Shanghai was superior to the national average. Especially in terms of BYO eco-friendly shopping bags and reuse of plastic bags, the environmental behavior of residents in Shanghai was commendable. For waste sorting, the survey data came from CGSS of 2013 when there were basically no sorting bins in communities, so Shanghai was inferior to the national average in terms of this item; however, after

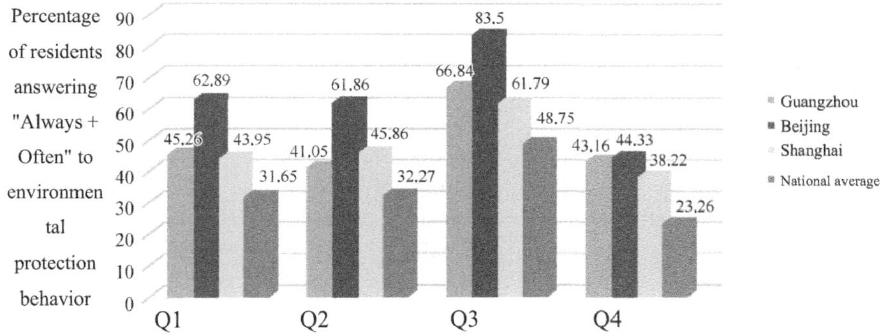

Fig. 11.2 Comparison of environmental awareness of residents in Beijing, Shanghai, Guangzhou and other parts of China

the implementation of the waste sorting policy in Shanghai, its data will have a significant advantage.

Besides, the percentages of residents in Beijing answering "Often" to the above three questions were respectively: 17.09% to waste sorting; 47.47% to BYO shopping basket or bag when shopping for daily necessities; 55.06% to the reuse of plastic bags. The three data values were all higher than the national average with regard to similar questions. The residents in Guangzhou were in a similar situation.

In terms of the comparison between the data of Shanghai and that of the rest of China, and the comparison between the three first-tier cities of Beijing, Shanghai and Guangzhou and the rest of China, we could find that in regions with relatively developed economy and more immigrants (implying that residents therein were higher-educated on the whole), residents had stronger environmental awareness, and more outstanding environmental behavior. This provided important evidence for the proposed progressive environmental "reform" in stages and batches.

2. Model estimation of residents' WTP for environmental protection in different regions

Table 11.6 compares the model estimation of Shanghai with that of the rest of China. The explained variable is "residents' happiness", similar to that of Chap. 8, and the graded data of general social science is used. As regards respondents' evaluation on their own happiness, "very unhappy" is denoted by 1; "relatively unhappy" is denoted by 2; "falling in between" is denoted by 3; "relatively happy" is denoted by 4; "completely happy" is denoted by 5. Through a regression analysis, we found that no matter for national residents or residents in Shanghai, NO_2 had a negative impact on happiness, and income had a positive impact thereon, which was consistent with the previous model conclusion. Residents' WTP could be measured by the aforesaid happiness measurement method.

Here, we need to pay attention to a set of data. In the comparison between residents in Shanghai and national residents, are they have the same WTP for environmental public goods? If different, who have a stronger WTP?

According to Eq. (8.4) in Chap. 8, we could estimate that in both Models 1 and 2, Shanghai residents' AWTP for environmental protection was higher than

Table 11.5 Statistics on environmental behavior of residents in Shanghai and other parts of China

Option	Waste sorting		BYO shopping basket or bag when shopping for daily necessities		Reuse plastic bags	
Comparison	National average	Shanghai	National average	Shanghai	National average	Shanghai
Never	55.12	53.69	24.17	21.37	18.72	16.03
Occasionally	32.41	38.68	35.14	32.32	31.78	26.97
Often	12.27	7.63	39.46	46.06	49.25	57.00

Data source Calculated according to the data of CGSS 2013
Unit: %

Table 11.6 Estimated results of basic models

Explained variable	Subjective happiness			
	Model 1 (Shanghai)	Model 2 (Shanghai)	Model 3 (national average)	Model 4 (national average)
NO_2	−11.69* (0.065)	−7.38 (0.028)	−2.23*** (0.000)	−1.254** (0.0610)
Log of household income	0.054*** (0.000)	0.056*** (0.014)	0.0626*** (0.000)	0.043*** (0.003)
Health condition				0.313*** (0.012)
Number of years of education received		−0.0084 (0.006)		0.0117*** (0.002)
Male				−0.060*** (0.012)
Urban residents				0.027** (0.014)
Observed value	1066	1057	25,655	19,414
R^2	0.014	0.035	0.043	0.053

Note *, ** and *** respectively stand for being significant at the levels of 10%, 5% and 1%; the number in brackets denotes standard deviation

RMB 3000; in Model 3, national residents' AWTP for environmental protection was 1472.70; in Model 4, the result was 1205.62. Through the comparison, we could find that Shanghai residents' AWTP for environmental protection was higher than the national average. Combining this with previous survey results, we could observe that Shanghai residents' environmental awareness and environmental action were both higher than the national average. This gave us the enlightenment that regions like Shanghai could take the lead in employing the mode of residents' participation in supply of public goods.

11.3.4 Progressive Guidance for Residents to Participate in Supply of Environmental Public Goods in the Light of Local Conditions

In the previous research herein, through model reasoning, we have drawn the following conclusions: higher income families have a stronger WTP for environmental protection; higher-educated residents are more willing to pay more fees for environmental protection; families with children are more willing to contribute to environmental protection; urban residents are more willing to pay pollution control fees.

1. Reason for the progressive "reform" of environmental public goods supply

In this chapter, we comparatively researched a series of data concerning the first-tier cities of Beijing, Shanghai and Guangzhou and the national average, including residents' environmental awareness, actual environmental action, and WTP for environmental protection. We could observe that first-tier city residents' support for environmental protection and governance is far stronger than the national average. Using the happiness measurement method in model estimations, we have found that Shanghai residents' WTP for environmental protection is far higher than the national average. From a national perspective, we will definitely achieve twofold results with half the effort by motivating residents to participate in environmental protection, but considering that China's economic development is imbalanced, with an obvious gap between coastal and inland regions, it is unrealistic to require those residents who are still endeavoring to catch up with a moderately prosperous society to give priority to paying for environmental protection.

2. Guidance for residents to participate in supply of environmental public goods in stages

In the report of the 19th CPC National Congress, Comrade Xi Jinping stressed that as socialism with Chinese characteristics has entered a new era, the principal contradiction in Chinese society has evolved into one between unbalanced and inadequate development and the people's ever-growing needs for a better life. On the one hand, it shows that Chinese residents yearn for a "better life" (including the pursuit of a high-quality environment); on the other hand, it emphasizes the imbalance of development between regions. Based on the reality, we suggest encouraging residents to voluntarily participate in supplying environmental public goods in stages. Economically developed regions may take the lead; the government may work out a schedule, promote the mode used in regions which have achieved significant results, introduce it to second and third-tier cities after achieving results in first-tier cities, and then introduce to all parts of China. This progressive reform is a nice choice for China at present.

For example, the Yangtze River Delta (hereinafter referred to as "YRD"), is an economic belt led by Shanghai and composed of Jiangsu, Zhejiang, Anhui, etc., China's first economic zone, the nationally-recognized strongest economic center in terms of comprehensive strength, an important international gateway in the Asian-Pacific region, a globally important advanced manufacturing base, as well as China's first world-class city cluster. As China's largest economic core region, the YRD may take the lead in adopting this mode.

The YRD covers an area of about 358,000 km^2, approximately accounting for 3.72% of the total area of China. Besides, it is also a part of the Middle-Lower Yangtze plains, with numerous rivers and lakes, mild and humid climate, abundant water resources, as well as favorable national conditions. However in recent years, with the rapid development of industrial and agricultural production, the surge in population and the increase in the consumption of fertilizers and pesticides, a large amount of insufficiently treated industrial wastewater and domestic sewage have been discharged into rivers, and polluted the environment. This has become a tricky problem faced by the YRD City Cluster. Furthermore, with a solid foundation in economic strength, the YRD is capable of playing a leading role in environmental governance.

Figure 11.3 shows the statistics on the regional GDP and PM_{10} of the YRD from 2004 to 2016. We could observe that from 2004 to 2016, the regional GDP of the YRD showed a rapid growth on the whole. However, the rapid economic growth was accompanied by increasingly severe environmental pollution. The deterioration of air quality was increasingly flat, but still showed an upward trend.

As regards the economic conditions of this region, the YRD has always been the most economically developed region in China. Figure 11.4 depicts the comparison between the per-capita GDP of the YRD and the national per-capita GDP. Seen from Fig. 11.4, the per-capita GDP of the YRD was higher than (nearly twice) the national average in most periods, except during the financial crisis of 2008. Its high economic level implies that the government and local residents are willing to invest more funds in providing a larger quantity of environmental public goods.

Table 11.7 lists the S&T indexes of Shanghai, Jiangsu, Zhejiang and the entire country in promoting socioeconomic development from 2008 to 2016. The table suggested that the national average maintained a year-on-year growth from 2008 to 2012, but had a downward trend from 2012 to 2013; the YRD's S&T index

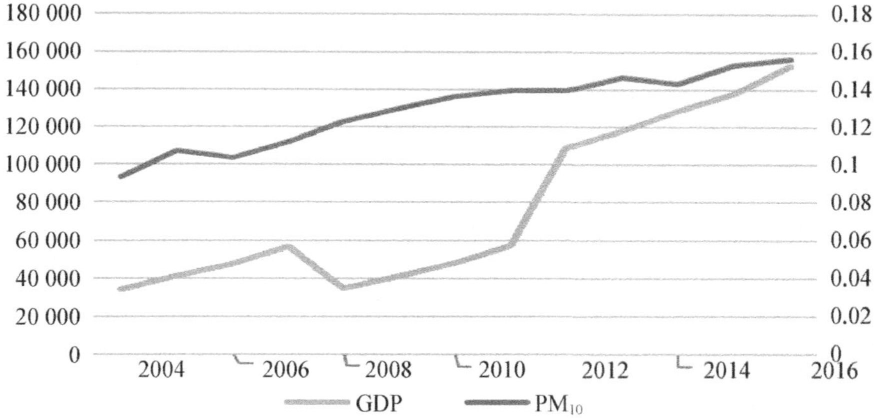

Fig. 11.3 Changes in regional GDP and PM_{10} of the YRD from 2004 to 2016. *Data source* The National Bureau of Statistics

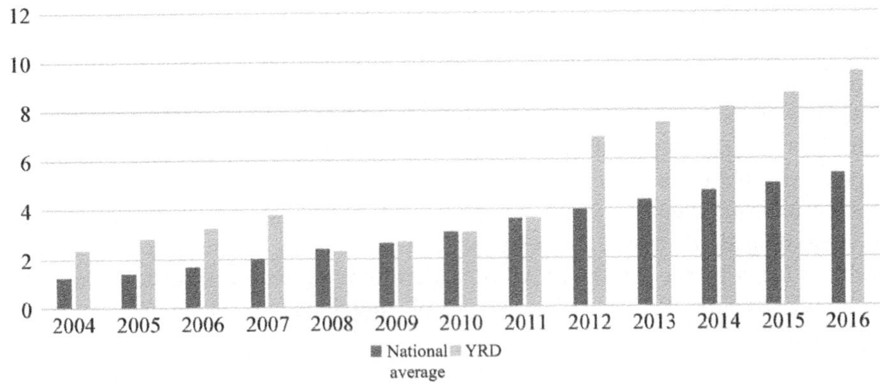

Fig. 11.4 Comparison between the per-capita GDP of the YRD and the National Per-Capita GDP. *Data source* The National Bureau of Statistics

in promoting socioeconomic development was higher than the national average. Shanghai had been among the top 3 nationwide; Zhejiang made significant progress, rose to the 4th place in 2014, and maintained the place for three years. Meanwhile, Zhejiang's S&T index had been relatively stable and ranked 5th to 7th. Therefore, the YRD leads the country in terms of S&T innovation. High S&T innovation level implies that the YRD has enough S&T innovation strength to provide higher-quality environmental public goods.

From the economic perspective, in the aspect of S&T in guiding economic progress, the above demonstrates that China's YRD is fit to take the lead in carrying

Table 11.7 S&T indexes of Shanghai, Jiangsu, Zhejiang and the entire country in promoting socioeconomic development

Year	S&T index in promoting socioeconomic development/%				National ranking for S&T in promoting economy		
	National average	Shanghai	Jiangsu	Zhejiang	Shanghai	Jiangsu	Zhejiang
2008	51.36	81.92	62.72	66.42	1	7	6
2009	57.02	82.20	70.70	70.96	2	7	6
2010	62.65	83.15	73.24	72.63	1	6	7
2011	65.19	83.50	74.59	73.64	1	6	7
2012	67.98	83.36	75.82	75.26	1	6	9
2013	62.78	82.25	68.26	69.09	1	6	8
2014	62.84	79.25	70.85	72.77	3	5	4
2015	64.99	77.11	69.66	72.75	3	6	4
2016	71.66	80.5	77.96	78.46	2	5	4

Data source Website of the China Science and Technology Statistics

out the supply "reform" of environmental public goods. In terms of promoting residents' participation in the supply-side reform of environmental public goods in batches and stages, reference can be made to the operation modes of the "Five Water Co-governance" in Zhejiang and the "Waste Sorting" in Shanghai; the government should play the leading role, adopt institutional guarantee, and motivate the entire people to actively participate in the supply of environmental public goods by various incentive measures.

Chapter 12
Main Conclusions and Prospects

12.1 Main Conclusions

This book extends the application of the public goods theory to environmental governance, analyzes residents' effective demand revelation and how to avoid the "free-rider" problem in the supply process in terms of the spatial and intergenerational externalities of environmental public goods, focuses on probing into the effective supply mechanism of environmental public goods based on intergenerational externality characteristics, measures residents' WTP for environmental protection by the happiness measurement method, and further presents the mechanism for government-led supply of environmental public goods by residents and enterprises through theoretical modeling, thereby providing a certain theoretical basis for the effective supply of environmental public goods. The following research conclusions are drawn herein.

1. Environmental public goods have the double attributes of spatial and intergenerational externalities, and the intergenerational externality leads to the serious undersupply thereof. The spillover of intergenerational costs or benefits arising from the intergenerational externality of environmental public goods will give rise to asymmetric costs and benefits between present and future generations, and the absence of the agent of future generations. Starting with the rational man hypothesis, the present generation makes decisions on their own side, with little consideration for the interests of future generations. As a result, compared to "Backward Intergenerational Public Goods" which provide the present generation old-age security, "Forward Intergenerational Public Goods" to the benefit of future generations are in seriously short supply.
2. China's existing supply mechanism of environmental public goods is incapable of solving the serous pollution problem. Through analyzing the status of air, water and waste pollution and making an international comparison, we found that environmental pollution has severely damaged our living environment, disturbed our lives, hindered our progress from material to spiritual life, greatly jeopardized

J. Yang, *Effective Supply Mechanism and Path of Environmental Public Goods in China*, Public Economy and Urban Governance in China,
https://doi.org/10.1007/978-981-99-2939-9_12

our physical health, and destroyed our human capital. China has collected pollution fees from polluters since 1979, implemented pollution governance for many years, and continued to increase investment in governance every year. But seen from the current situation, the effect is insignificant. In addition to the generally mentioned government failure and market failure, according to analytic findings, one big problem lies in the fact that the supply mechanism of environmental public goods fails to function effectively. Thus, to establish an effective supply mechanism is a matter of great urgency.

3. Motivating residents to participate in environmental governance based on an intergenerational perspective can bring Pareto improvement. When analyzing all sorts of pollution sources, we have found that not only industrial enterprises are the chief culprit of environmental pollution, because their emitted NO_x, SO_2 and soot (dust) have put us in a terrible environment; domestic wastewater discharged by residents has shown a rising trend year by year, and household waste pollution has greatly stunned us. If it is possible to guide residents to join the environmental governance team, on the one hand, their environmental awareness can be elevated, and environmental pollution can be minimized; on the other hand, the shortage of funds in pollution control can be alleviated.
4. The current material life conditions and residents' environmental awareness can support the shift of residents as the "demand side" of clean environment to the "supply side" of quality environment.

   Through an international comparison of per-capita GDP and residents' Engel's coefficient in combination with an analysis on the economic take-off stages proposed by Rostow, we could find: Chinese residents' material life conditions have reached moderate prosperity, and most residents have begun to "search for quality", i.e. reaching the stage dominated by the service sector and the environmental renovation sector as Rostow mentioned, where residents may attach increasing importance to outdoor consumption, and show increasing concern for the ecological environment.

   Additionally, the CGSS data suggests that if the government takes the lead in taking measures to allow residents to take environmental protection action in a more convenient manner, residents will have a high probability to invest in environmental protection. According to the survey results of such items as BYO eco-friendly shopping bags and reuse of shopping bags which residents can achieve objectively, a high percentage of residents may take environmental protection action.
5. The life satisfaction measurement method can be used to measure residents WTP for environmental protection. Through a comparison between the methods of public goods valuation—the revealed preference method and the stated preference method, academic researchers have found another method of measuring environmental public goods—happiness measurement method. The method associates the positive effect (the increase of income enhances happiness) with the negative effect (environmental pollution lowers residents' happiness) of economic growth for research, thereby being capable of measuring the amount of money that residents are willing pay for lowering every unit of pollutant (or

increasing a day with air quality above level II every year). This is the huge progress in academic research.

6. The mechanism of government-led environmental governance with the joint participation of residents and enterprises based on the OLG model can bring multiple dividends to society. It is a new trend to probe into environmental governance with residents as the "supply side" of quality environment. When our material standard of living reaches a certain level, residents will change their ideology. Environmental governance is an intergenerational social issue. If the government's "intergenerational contract" is rationally designed, and the happiness measurement method is employed to measure residents' MWTP for environmental governance in different regions, the policy implementation will not only improve environmental quality and accelerate economic growth, but also increase the human capital of future generations. The theoretical model fully demonstrates that the cooperation among the government, enterprises and residents in environmental governance can bring dividends.
7. The "intergenerational contract" used by a long-term government can ensure the effective functioning of the mechanism of government-led environmental governance with the joint participation of residents and enterprises. The effective functioning of the mechanism depends on supporting policies. The government maintains principal-agent relations with enterprises, residents, and the environmental protection fund company respectively, and the "intergenerational contract" based on the incomplete contract theory needs a credible institutional guarantee, thus the government' credibility is particularly important. Besides, the long-term government and the assessment indicators for the government are crucial, and need to be intensified in the policymaking process.
8. Residents' participation in the supply of environmental public goods can bring better performance. The "Five Water Co-governance" in Zhejiang and the pilot household waste sorting in Shanghai both confirm the saying, "When everybody adds fuel, the flames rise high." As long as the government gives proper guidance, and matches rewards & punishments with incentives, the intervention of nongovernmental forces will achieve twofold results with half the effort in the supply of public goods.
9. Comparatively speaking, the progressive reform is more appropriate for China's supply of public goods. In consideration of China's regional gaps mainly resulting from the imbalance of economic development, including the north–south gap and the gap between coastal and inland regions, it is nearly infeasible to implement "one-size-fits-all" policies in China. We suggest progressively providing environmental public goods in stages based on the economic development status, residents' income, educational attainment, and pursuit of environmental quality, etc. in different regions.

12.2 Deficiencies and Prospects

Though this book uses a large quantity of micro-data for analysis, and draws relatively robust empirical conclusions, it still has lots of deficiencies and plenty of room for improvement.

First, some of the data is imperfect. Considering limited information, this study uses cross-section sample survey data. The sample data size has no uniform standard, and research conclusions are based on the survey sample analysis. Data collection and improvement need to be enhanced in the future.

Second, the supply mechanism of environmental public goods fails to be discussed in details. Although this study is focused on the analysis from the intergenerational perspective, as regards the future research orientation, the specific operation mechanism for connecting supply with demand may be designed in accordance with different types of environmental public goods in different regions.

Third, the happiness measurement method has limitations in practical application. The method is capable of revealing residents' demand preference for environmental public goods, but there is a certain gap in the material living standards in all parts of China, and residents' assessment on happiness is subjective. As a result, in case of using this method to measure residents' WTP for environmental protection, we need to conduct separate measurements according to the data of different regions, and have difficulty in applying a uniform standard. Certainly, measurements and calculation may be performed according to the regional division of the YRD, the Beijing-Tianjin-Hebei region, and the Pearl River Delta. This will be an orientation of our future search.

References

1. Chen Dedi, Li Zhou, Ku Guisheng. 2001. National defense economy dictionary. Beijing: Military Science Publishing House.
2. van Praag, B.M.S., and B.E. Baarsma. 2005. Using happiness surveys to value intangibles: The case of airport noise. *The Economic Journal* 115 (500): 224–246.
3. Almond, D., Y. Chen, M. Greenstone, et al. 2009. Winter heating or clean air? Unintended impacts of China's Huai river policy. *The American Economic Review* 99 (2): 184–190.
4. Hanna, R., and P. Oliva. 2015. The effect of pollution on labor supply: evidence from a natural experiment in Mexico City. *Journal of Public Economics* 122: 68–79.
5. Chen, Y., A. Ebenstein, M. Greenstone, et al. 2013. Evidence on the impact of sustained exposure to air pollution on life expectancy from China's Huai River policy. *Proceedings of the National Academy of Sciences of the United States of America* 110 (32): 12936–12941.
6. Shuo, Chen, and Chen Ting. 2014. Air pollution and public health: evidence from sulfur doxide emission of coal-fired power stations in China [J]. *Economic Research* 8: 158–169.
7. Sandler, T., and V.K. Smith. 1976. Intertemporal and intergenerational Pareto efficiency. *Journal of Environmental Economics and Management* 2 (3): 151–159.
8. Sandler, T. 1982. A theory of intergenerational clubs. *Economic Inquiry* 20 (2): 191–208.
9. Shiliang, Zhao. 2003. Externality between generations and the cause of unsustainable development. *China Population, Resource and Environment* 13 (4): 1–4.
10. Haijing, Sun. 2013. On intergenerational compensation of intergenerational public goods. *Journal of Shandong Institute of Business and Technology* 27 (4): 23–26.
11. Haijing, Sun. 2012. Study on environmental regulation of intergenerational supply of public goods perspective. *Economy and Management* 5 (26): 17–20.
12. John, A., and R. Pecchenino. 1994. An overlapping generations model of growth and the environment. *The Economic Journal* 104 (427): 1393–1410.
13. John, A., R. Pecchenino, D. Schimmelpfennig, et al. 1995. Short-lived agents and the long-lived environment. *Journal of Public Economics* 58 (1): 127–141.
14. Diamond, P.A. 1965. National debt in a neoclassical growth model. *The American Economic Review* 55 (5): 1126–1150.
15. Ostrom, E. 1990. *Governing the commons: The evolution of institutions for collective action.* Cambridge: Cambridge University Press.
16. Olson, M. 1971. *The logic of collective action.* Cambridge, MA: Harvard University Press.
17. Tiebout, C.M. 1956. A pure theory of local expenditures. *Journal of Political Economy* 64 (5): 416–424.
18. Yang Jibo, Wu Baijun. 2015. Review on the study of intergenerational public goods: the connotation, path and the new perspective of governance. *Journal of East China University of Science and Technology (Social Science Edition)* (3):60–68.

J. Yang, *Effective Supply Mechanism and Path of Environmental Public Goods in China*, Public Economy and Urban Governance in China,
https://doi.org/10.1007/978-981-99-2939-9

19. Rangel, A. 2003. Forward and backward intergenerational goods: Why is social security good for the environment? *American Economic Review* 93 (3): 813–834.
20. Welsch H, Kühling J. 2009. Using happiness data for environmental valuation: issues and applications. *Journal of Economic Surveys* 23(2):385–406.
21. Welsch, H. 2002. Preferences over prosperity and pollution: Environmental valuation based on happiness surveys. *Kyklos* 55 (4): 473–494.
22. Levinson, A. 2012. Valuing public goods using happiness data: The case of air quality. *Journal of Public Economics* 96 (9–10): 869–880.
23. Samuelson, P.A. 1954. The pure theory of public expenditure. *The Review of Economics and Statistics* 36 (4): 387–389.
24. Xiaoli, Xiong. 2010. *A Study on the Transfer of Environment Welfare among Generations.* Nanchang: Jiangxi University of Finance and Economics.
25. Cornes, R., and T. Sandler. 1996. *The theory of externalities, public goods, and club goods.* Cambridge: Cambridge University Press.
26. Lowry, W.R. 1998. *Preserving public lands for the future: The politics of intergenerational goods.* Washington DC: Georgetown University Press.
27. Li Yufang, Sun Haijing. 2009. Intergenerational equity and supply of intergenerational public goods. *Social Sciences in Guangdong* (3):14–19.
28. Musgrave, R.A. 1939. The voluntary exchange theory of public economy. *The Quarterly Journal of Economics* 53 (2): 213–237.
29. Samuelson, P.A. 1955. Diagrammatic exposition of a theory of public expenditure. *Review of Economics and Statistics* 21: 350–356.
30. Dawes, R.M., and B. Corrigan. 1974. Linear models in decision making. *Psychological Bulletin* 81 (2): 95–106.
31. Dawes R.M. 1975. 4 formal models of dilemmas in social decision-making. University of Oregon and Oregon Research Institute.
32. Hardin, G. 1968. The tragedy of the commons. *Science* 162 (3859): 1243–1248.
33. Bergstrom, T., L. Blume, and H. Varian. 1986. On the private provision of public goods. *Journal of Public Economics* 29 (1): 25–49.
34. Liming, Fan. 2005. *Analysis on China's Public Goods Market and Voluntary Provision.* Shanghai: Shanghai People's Publishing House.
35. Becker, G.S. 1974. A theory of social interactions. *Journal of Political Economy* 82 (6): 1063–1093.
36. Becker G.S., Tomes N. 1976. Child endowments and the quantity and quality of children. *Journal of Political Economy* 84(4, Part 2): 143–S162.
37. Collard, D.A. 1978. *Altruism and economy: A study in non-selfish economics.* Oxford: Oxford University Press.
38. Andreoni, J. 1995. Warm-glow versus cold-prickle: The effects of positive and negative framing on cooperation in experiments. *The Quarterly Journal of Economics* 110 (1): 1–21.
39. Simon H.A. 2002. Selections of Simon. Huang Tao, Trans. Beijing: Capital University of Economics and Business Press.
40. Hodgson, G.M. 1997. The ubiquity of habits and rules. *Cambridge Journal of Economics* 21 (6): 663–684.
41. Andreoni, J. 1990. Impure altruism and donations to public goods: A theory of warm-glow giving. *The Economic Journal* 100 (401): 464–477.
42. Hu Shiqing, Wu Jiapei. 2009. From Altruism to social rationality -- an integrated view of economics research on Altruism. *Research on Financial and Economic Issues* (6):3–10.
43. Ye Hang. 2005. An explanation of economics to Altruistic behavior. *Economist* 3:22–29.
44. Hamilton W.D. 1964. The genetical evolution of social behaviour. II. *Journal of Theoretical Biology* 7(1):17–52.
45. Trivers R.L. 1971. The evolution of reciprocal altruism. *The Quarterly Review of Biology* 46(1):35–57.
46. Axelrod, R., and W.D. Hamilton. 1981. The evolution of cooperation. *Science* 211 (4489): 1390–1396.

47. Rabin, M. 1993. Incorporating fairness into game theory and economics. *The American Economic Review* 83 (5): 1281–1302.
48. Fehr, E., and K.M. Schmidt. 1999. A theory of fairness, competition, and cooperation. *The Quarterly Journal of Economics* 114 (3): 817–868.
49. Bolton, G.E., and A. Ockenfels. 2000. ERC: a theory of equity, reciprocity, and competition. *American Economic Review* 90 (1): 166–193.
50. Edwards, W. 1962. Utility, subjective probability, their interaction, and variance preferences. *Journal of Conflict Resolution* 6 (1): 42–51.
51. Wilson, R. 1969. An axiomatic model of logrolling. *American Economic Review* 59 (3): 331–341.
52. Ross, S.A. 1973. The economic theory of agency: The principal's problem. *American Economic Review* 63 (2): 134–139.
53. Mirrless, J. 1976. The optimal structure of incentives and authority within an organization. *Bell Journal of Economics* 7 (1): 105–131.
54. Arrow, K.J. 2001. Uncertainty and the welfare economics of medical care. *Journal of Health Politics, Policy and Law* 26 (5): 851–883.
55. Holmstrom, B. 1979. Moral hazard and observability. *Bell Journal of Economics* 10 (1): 74–91.
56. Grossman, S.J., and O.D. Hart. 1983. An analysis of the princinpal-agent problem. *Econometrica* 51 (1): 302–340.
57. Williamson, O.E. 2002. The theory of the firm as governance structure: From choice to contract. *Journal of Economic Perspectives* 16 (3): 171–195.
58. Coase, R.H. 1937. The nature of the firm. *Economica* 4 (16): 386–405.
59. Williamson O.E. 1985. The economic institutions of capitalism: firms, markets, relational contracting. New York: the Free Press.
60. Dahlman, C.J. 1979. The problem of externality. *The Journal of Law and Economics* 22 (1): 141–162.
61. Hart, O., and J. Moore. 2008. Contracts as reference points. *The Quarterly Journal of Economics* 123 (1): 1–48.
62. Bolton P, Dewatripont M. 2005. Contract theory. Cambridge: the Massachusetts Institute of Technology Press.
63. Gibbons, R. 2005. Four formal (izable) theories of the firm? *Journal of Economic Behavior & Organization* 58 (2): 200–245.
64. Shicheng, Jiang, and Fei Fangyu. 2008. from ex ante inefficiency to ex post inefficiency: Comparing several classical types of models of incomplete contract theory. *Economic Research* 8: 145–156.
65. Jiang Shicheng, Li Jing, Fei Fangyu. 2018. Research progress of endogenous incomplete contract theory. *Economic Perspectives* (5):102–116.
66. Li Yufang. 2005. An analysis on externalities of government supply behavior of public goods. *South China Journal of Economics* 6:21–23.
67. Xiao, Guo, and Xia Hongsheng. 2006. Theoretical research about the effective way of solving intergenerational externality. *China Industrial Economy* 12: 62–68.
68. Li Yufang, Sun Haijing. 2009. Analysis on the mechanism for effect of local government competition on intergenerational public goods supply. *Academic Exchange* (1):58–63
69. Zhongjie, Guo. 2006. *Intergenerational Externality and Sustainable Development*. Guangzhou: Ji'nan University.
70. Weiss, E.B. 1984. The planetary trust: conservation and intergenerational equity. *Ecology Law Quarterly* 19 (3): 284–286.
71. Page, T. 1977. *Conservation and economic efficiency: an approach to materials policy*. Washington D. C: The Johns Hopkins University Press.
72. Pearce, D. 1996. *Blueprint for a green economy*. Beijing: Beijing Normal University Press.
73. Weiss, E.B. 1990. Our rights and obligations to future generations for the environment. *American Journal of International Law* 84 (1): 198–207.

74. Ling, Wei, and Ren Baoping. 2002. Thoughts on the relation between external effects and sustained development. *Contemporary Economic Research* 6: 7–8.
75. Doeleman J.A, Sandler T. 1998. The intergenerational case of missing markets and missing voters. *Land Economics* 1–15.
76. Kotlikoff, L.J., and R.W. Rosenthal. 1993. Some inefficiency implications of generational politics and exchange. *Economics & Politics* 5 (1): 27–42.
77. Oates, W.E. 1973. The effects of property taxes and local public spending on property values: a reply and yet further results. *Journal of Political Economy* 81 (4): 1004–1008.
78. Atkinson A.B, Stiglitz J.E. 1976. The design of tax structure: direct versus indirect taxation. *Journal of Public Economics* 6(1–2):55–75.
79. Warr, P.G. 1983. The private provision of a public good is independent of the distribution of income. *Economics Letters* 13 (2–3): 207–211.
80. Belk, R.W. 1975. Situational variables and consumer behavior. *Journal of Consumer Research* 2 (3): 157–164.
81. Kreps, D.M., and R. Wilson. 1982. Reputation and imperfect information. *Journal of Economic Theory* 27 (2): 253–279.
82. Warr, P.G. 1982. Pareto optimal redistribution and private charity. *Journal of Public Economics* 19 (1): 131–138.
83. Webster Jr C.E. 1983. The effects of deficits on interest rates. *Economic Review* 19–28.
84. Isaac, R.M., J.M. Walker, and S.H. Thomas. 1984. Divergent evidence on free riding: An experimental examination of possible explanations. *Public Choice* 43 (2): 113–149.
85. Sugden, R. 1984. Reciprocity: The supply of public goods through voluntary contributions. *The Economic Journal* 94 (376): 772–787.
86. Andreoni, J. 1988. Privately provided public goods in a large economy: The limits of altruism. *Journal of Public Economics* 35 (1): 57–73.
87. Isaac, R.M., J.M. Walker, and S.H. Thomas. 1984. Divergent evidence on free riding: An experimental examination of possible explanations [J]. *Public Choice* 43 (2): 113–149.
88. Ostrom, E. 2000. Collective action and the evolution of social norms. *Journal of Economic Perspectives* 14 (3): 137–158.
89. Xi Heng. 2003. Study on the supply mechanism of public goods, Xi'an: Northwest University.
90. Laffont, J.J. 1987. Incentives and the allocation of public goods. *Handbook of Public Economics* 2 (2): 395–396.
91. Bowles, S., and S.H. Hwang. 2008. Social preferences and public economics: Mechanism design when social preferences depend on incentives. *Journal of Public Economics* 92 (8–9): 1811–1820.
92. He Yanling. 2001. Diversification of third sector and community supply channels of public goods. *Public Management Research* 240–253.
93. Boadway, R. 2007. Policy forum: the annual tax expenditure accounts-a critique. *Canadian Tax Journal* 55 (1): 106.
94. Yean, Zhou, Lian Hongquan, Chen Yefeng, et al. 2013. Social role, heterogenous preferences and public goods provision. *Economic Research* 1: 124–137.
95. Hongquan, Lian. 2014. Punishment and social cooperation: a selective survey and discussion on experimental economics. *South China Journal of Economics* 9: 128–134.
96. Yan, Liu. 2015. Motivations of intergenerational transfers: altruism or exchange – evidence from CHARLS. *Economic Theory and Business Management* 298 (10): 58–68.
97. Coates, D. 1996. A diagrammatic demonstration of public crowding-out of private contributions to public goods. *The Journal of Economic Education* 27 (1): 49–58.
98. Ahmed, S.A., and M. Ali. 2004. Partnerships for solid waste management in developing countries: Linking theories to realities. *Habitat International* 28 (3): 467–479.
99. Pigou, A.C. 1928. An analysis of supply. *The Economic Journal* 38 (150): 238–257.
100. Dye, T.R. 1966. *Politics, economics, and the public: Policy outcomes in the American states.* Chicago: Rand McNally.
101. Kneese A.V., Bower B.T. 1979. Environmental quality and residuals management [EB/OL]. 01 01 1979. http: //www. osti.gov/biblio/6192188.

102. Pearce, D. 1991. The role of carbon taxes in adjusting to global warming. *The Economic Journal* 101 (407): 938–948.
103. Bovenberg, A.L., and R.A. De Mooij. 1994. Environmental levies and distortionary taxation. *The American Economic Review* 84 (4): 1085–1089.
104. Bovenberg, A., and L.H. Goulder. 1996. Optimal environmental taxation in the presence of other taxes: General-equilibrium analyses. *The American Economic Review* 86 (4): 985–1000.
105. de Mooij R, Bovenberg L. 1997. Environmental tax reform and endogenous growth. *Journal of Public Economics* 207–237.
106. Goulder, L.H. 1998. Environmental policy making in a second-best setting. *Journal of Applied Economics* 1 (2): 279–328.
107. Fullerton, D., and S.R. Kim. 2008. Environmental investment and policy with distortionary taxes, and endogenous growth. *Journal of Environmental Economics and Management* 56 (2): 141–154.
108. Li Qiyun, Zong Bin, Li Zhengyu. 2007. Optimal environmental tax: pigovian rule and tax coordination. *China Population, Resources and Environment* 17(6):18–22.
109. Fengliang, Liu, and Lyu Zhihua. 2009. Research on optimal environmental tax and its supporting policies in the framework of economic growth – simulation based on Chinese data. *Journal of Management World* 6: 47–58.
110. Dongmin, Zhang, and Jin Chengxiao. 2014. Pollution tax, accumulation of healthy human capital and long-run economic growth. *Finance & Economics* 321 (12): 79–87.
111. Xiao Xinrong, Liao Pu. 2014. Research on optimal government investment in pollution control. *Journal of World Economy* (1):106–119.
112. Greenstone, M., and R. Hanna. 2014. Environmental regulations, air and water pollution, and infant mortality in India. *American Economic Review* 104 (10): 3038–3072.
113. Requate, T., and W. Unold. 2003. Environmental policy incentives to adopt advanced abatement technology: Will the true ranking please stand up? *European Economic Review* 47 (1): 125–146.
114. Cherry, T.L., S. Kallbekken, and S. Kroll. 2012. The acceptability of efficiency-enhancing environmental taxes, subsidies and regulation: An experimental investigation. *Environmental Science & Policy* 16: 90–96.
115. He Jianwu, Li Shantong. 2009. The impact of energy-saving and emission-reduction environmental tax on China's economy. *Journal of Quantitative & Technical Economics* (1):32–45.
116. Liu Xiaobing. 2004. Experimental research on individual cooperation in providing public goods. *Journal of Management World* (2):51–56, 97.
117. Yingna, Huang, Guo Zhenren, Zhang Tianzhu, et al. 2005. Quantitative Analysis on the Feasibility of Energy-Environment Tax Implementation in China with a CGE Model [J]. *Urban Environment & Urban Ecology* 18 (2): 18–20.
118. Li Dongdong, Yang Jingyu. 2015. R&D subsidy and environmental tax combination based on growth model. *Studies in Science of Science* 33(7):1026–1034.
119. Yuanpeng, Zhang, and Lin Dawei. 2015. Social preferences, reward & punishment mechanism and public goods provision – research based on an experimental method. *South China Journal of Economics* 33 (12): 26–39.
120. Fudenberg, D., D.M. Kreps, and E.S. Maskin. 1990. Repeated games with long-run and short-run players. *The Review of Economic Studies* 57 (4): 555–573.
121. Ono, T. 1996. Optimal tax schemes and the environmental externality. *Economics Letters* 53 (3): 283–289.
122. Rangel, A. 2005. How to protect future generations using tax-base restrictions. *American Economic Review* 95 (1): 314–346.
123. Ono, T. 2008. Environmental tax reform in an overlapping-generations economy with involuntary unemployment. *Environmental Economics and Policy Studies* 9 (4): 213–238.
124. Jidong, Yang, and Zhang Yiran. 2014. Pricing of air pollution: analysis based on happiness data. *Journal of World Economy* 12: 162–188.

125. Qi Yu, Lu Hongyou. 2015. Crisis or turning point: the national health in economic cycle. *China Soft Science* (12):36–48.
126. Xiao, Zhang. 2014. Trend of and the governance system for water pollution in China. *China Soft Science* 10: 11–24.
127. Savas E.S. 1992. Privatization in post-socialist countries. *Public Administration Review* 573–581.
128. Hongye, Gao. 2014. *Western Economics (Macroeconomics)*, 6th ed. Beijing: China Renmin University Press.
129. Lavy V, Ebenstein A, Roth S. 2014. The impact of short term exposure to ambient air pollution on cognitive performance and human capital formation. National Bureau of Economic Research.
130. Chay, K.Y., and M. Greenstone. 2003. The impact of air pollution on infant mortality: Evidence from geographic variation in pollution shocks induced by a recession. *The Quarterly Journal of Economics* 118 (3): 1121–1167.
131. Luechinger, S. 2014. Air pollution and infant mortality: A natural experiment from power plant desulfurization. *Journal of Health Economics* 37: 219–231.
132. Schwartz, J., and A. Marcus. 1990. Mortality and air-pollution in London: A time-series analysis. *American Journal of Epidemiology* 131 (1): 185–194.
133. Mendelsohn, R., and G. Orcutt. 1979. An empirical analysis of air pollution dose-response curves. *Journal of Environmental Economics and Management* 6 (2): 85–106.
134. Barr, S. 2007. Factors influencing environmental attitudes and behaviors: A UK case study of household waste management. *Environment and Behavior* 39 (4): 435–473.
135. Briscoe, J., P.F. de Castro, C. Griffin, et al. 1990. Toward equitable and sustainable rural water supplies: A contingent valuation study in Brazil. *The World Bank Economic Review* 4 (2): 115–134.
136. Yan, Wang, Zhang Yisheng, and Liu Bin. 2008. Willingness to pay for reducing air pollution related diseases in Qingdao. *Journal of Environment and Health* 25 (4): 326–330.
137. Clarke, P.M. 1998. Cost-benefit analysis and mammographic screening: A travel cost approach. *Journal of Health Economics* 17 (6): 767–787.
138. Feng Hao, Lu Ming. 2010. Buying homes for schools: experimental evidence and policy implication of education impacting housing price. *Journal of World Economy* 6(12):89–104.
139. Zeng Xiangang, Xie Fang, Zong Quan. 2015. Behavior selection and willingness to pay for reducing $PM_{2.5}$ health risk: taking residents in Beijing as an example. *China Population, Resource and Environment* 25(1):127–133.
140. Li Yiran, Dong Lei, Meng Yilin, et al. 2014. Study on air pollution in Chengdu—survey on WTP of residents in Chengdu. *The Fortune Times* (6):29–30.
141. Tongyang, Wei, Jin Leshan, Jin Zongzhen, et al. 2015. Analysis on residents' WTP for air quality improvement in Beijing. *Urban Problems* 1: 75–81.
142. Easterlin, R.A. 1974. *Does economic growth improve the human lot? Some empirical evidence // Nations and households in economic growth*, 89–125. Pittsburgh: Academic Press.
143. Ferrer-I-Carbonell, A., and P. Frijters. 2004. How important is methodology for the estimates of the determinants of happiness? *The Economic Journal* 114 (497): 641–659.
144. Di Tella, R., and R. MacCulloch. 2008. Gross national happiness as an answer to the Easterlin Paradox? *Journal of Development Economics* 86 (1): 22–42.
145. Kahneman, D., and A. Tversky. 2003. Experienced utility and objective happiness: A moment-based approach. *The Psychology of Economic Decisions* 1: 187–208.
146. Kahneman, D., and R. Sugden. 2005. Experienced utility as a standard of policy evaluation. *Environmental and Resource Economics* 32 (1): 161–181.
147. Ng, Y.K. 1997. A case for happiness, cardinalism and interpersonal comparability. *The Economic Journal* 107 (445): 1848–1858.
148. Greenstone, M., and B.K. Jack. 2015. Envirodevonomics: A research agenda for an emerging field. *Journal of Economic Literature* 53 (1): 5–42.
149. Welsch, H. 2006. Environment and happiness: Valuation of air pollution using life satisfaction data. *Ecological Economics* 58 (4): 801–813.

150. Welsch, H., and J. Kühling. 2009. Using happiness data for environmental valuation: Issues and applications. *Journal of Economic Surveys* 23 (2): 385–406.
151. Luechinger, S., and P.A. Raschky. 2009. Valuing flood disasters using the life satisfaction approach. *Journal of Public Economics* 93 (3–4): 620–633.
152. Luechinger, S. 2010. Life satisfaction and transboundary air pollution. *Economics Letters* 107 (1): 4–6.
153. Frey, B.S., S. Luechinger, and A. Stutzer. 2009. The life satisfaction approach to valuing public goods: The case of terrorism. *Public Choice* 138 (3–4): 317–345.
154. Jidong, Yang, and Zhang Yiran. 2014. Pricing of air pollution: analysis based on happiness data. *Journal of World Economy* 12: 162–188.
155. Carlsson, F., and O. Johansson-Stenman. 2000. Willingness to pay for improved air quality in Sweden. *Applied Economics* 32 (6): 661–669.
156. Xincai, Gao, Yue Li, and Zhang Qinzhi. 2011. Analysis on factors Influencing WTP for air pollution in Lanzhou. *Urban Problems* 1: 62–65.
157. Huang Yongming, He Lingyun. 2013. Urbanization, environmental pollution and subjective well-being: an empirical study on China. *China Soft Science* (12):82–93.
158. Li Ying, Bai Mo, Zhang Wei, et al. 2002. Analysis on the influence factors of residents' Willingness to pay for improving air quality in Beijing. *China Population, Resource and Environment* 12(6):123–126.
159. Zheng, S., C. Sun, Y. Qi, et al. 2014. The evolving geography of China's industrial production: Implications for pollution dynamics and urban quality of life. *Journal of Economic Surveys* 28 (4): 709–724.
160. Mu Huaizhong, Fan Hongmin. 2014. Expansion of urbanization and residents' WTP for air pollution control. *Journal of Chinese Academy of Governance* (6):81–85.
161. Shuijun, Peng, and Bao Qun. 2006. Economic growth and environmental pollution: an empirical test for the environmental Kuznets Curve Hypothesis in China. *Research on Financial and Economic Issues* 8: 5–19.
162. Lianwei, Zhang, and Zhang Lin. 2017. The ecological environment succession and management of Yongding River in Beijing. *Journal of Beijing Union University (Humanities and Social Sciences)* 1: 124–130.
163. Rostow. 2010. The Stages of economic growth. Guo Xibao, Wang Songmao, Trans. Beijing: China Social Sciences Press.
164. Gerlagh, R. 2007. Measuring the value of induced technological change. *Energy Policy* 35 (11): 5287–5297.
165. Poortinga, W., L. Steg, C. Vlek, et al. 2003. Household preferences for energy-saving measures: a conjoint analysis. *Journal of Economic Psychology* 24 (1): 49–64.
166. Li Ying. 2008. Research on nongovernmental supply of urban forest ecological service based on residents' Willingness to Pay [D]. Harbin: Northeast Forestry University.
167. Yunmeng, Gao. 2016. An empirical analysis on the influencing factors of willingness to pay for environmental protection. *Sichuan Environment* 2: 130–133.
168. Chay, K.Y., and M. Greenstone. 2005. Does air quality matter? evidence from the housing market. *Journal of Political Economy* 113 (2): 376–424.
169. Currie, J., and M. Neidell. 2005. Air pollution and infant health: What can we learn from California's recent experience? *The Quarterly Journal of Economics* 120 (3): 1003–1030.
170. Becker, G.S. 1981. Altruism in the family and selfishness in the market place. *Economica* 48 (189): 1–15.
171. Junxue, Jia, Guo Qingwang, and Ning Jing. 2011. Traditional cultural belief, social security and economic growth. *Journal of World Economy* 8: 3–18.
172. Berle, A.A., and G.C. Means. 1933. *The modern corporation and private property [M]*. London: Transaction Publishers.
173. Jensen, M.C., and W.H. Meckling. 1976. Theory of the firm: managerial behavior, agency costs and ownership structure. *Journal of Financial Economics* 3 (4): 305–360.
174. Arrow, K.J. 1985. Informational structure of the firm. *The American Economic Review* 75 (2): 303–307.

175. Radner R. 1981. Monitoring cooperative agreements in a repeated principal-agent relationship. *Econometrica: Journal of the Econometric Society* 1127–1148.
176. Fama, E.F. 1980. Agency problems and the theory of the firm. *Journal of Political Economy* 88 (2): 288–307.
177. Holmstrom, B., and J.R.I. Costa. 1986. Managerial incentives and capital management. *The Quarterly Journal of Economics* 101 (4): 835–860.
178. McAfee R.P, McMillan J. 1991. Optimal contracts for teams. *International Economic Review* 561–577.
179. Solow, R.M. 1979. Another possible source of wage stickiness. *Journal of Macroeconomics* 1 (1): 79–82.
180. Shapiro, C., and J.E. Stiglitz. 1984. Equilibrium unemployment as a worker discipline device. *The American Economic Review* 74 (3): 433–444.
181. Zhang Weiying, Yu Hui. 1994. Evolution and latest developments of western enterprise theory. *Economic Research* 11:70–81.
182. Zhang Weiying. 1995. Principal-agent relations in public-owned economy: theoretical analysis and policy implications. Wu Youchang, Ma Jie, Trans *Economic Research* 4:10–20.
183. Pindyck, Rubinfeld. 2009. *Microeconomics*. Beijing: China Renmin University Press.
184. Beibei, Shi, Feng Chen, Zhang Yan, et al. 2017. Increasing marginal utility of environmental regulation dividends. *China Industrial Economy* 12: 40–58.
185. Kotlikoff, L.J., T. Persson, and L.E.O. Svensson. 1988. Social contracts as assets: a possible solution to the time-consistency problem. *American Economic Review* 78 (5): 662–677.
186. Bardhan, P.K., and D. Mookherjee. 2000. Capture and governance at local and national levels. *American Economic Review* 90 (2): 135–139.
187. Blanchard, O., and A. Shleifer. 2001. Federalism with and without political centralization: China versus Russia. *IMF Staff Papers* 48 (1): 171–179.
188. Chen Kang, Hillman A.L., Gu Qingyang. 2002. Fiscal recentralization and behavioral change of local governments: from the helping hand to the grabbing hand. *China Economic Quarterly* 2(1):111–130.
189. Fu Yong, Zhang Yan. 2007. Chinese-style decentralization and fiscal expenditure structure bias: the price of competition for growth. *Journal of Management World* (3):4–12.
190. Li Guoping, Wang Yiqi. 2016. Tragedy of the commons' theory and empirical study in the trans-boundary water pollution treatment. *Soft Science* 30(11):24–28.
191. Shihong, Zeng, and Xia Jiechang. 2015. Tragedy of the commons, transaction cost and haze governance – system thinking for effective supply of environmental technology service. *Research on Financial and Economic Issues* 1: 10–15.
192. Li Shasha, Niu Li. 2014. An analysis of the impact of technology advance on carbon dioxide emission in China -- based on the static and dynamic panel data models. *Research on Economics and Management* 10: 9–26.
193. Shiyi, Chen. 2011. Fluctuating decline model of China's carbon intensity and its economic interpretation. *Journal of World Economy* 4: 124–143.
194. Shihong, Zeng, and Wang Xiaoyan. 2014. Environmental policy tools and technology absorption incentive – heterogeneity, adaptability and coordination. *Review of Industrial Economics* 13 (1): 105–118.
195. Xiaofu, Hou. 2004. Public participation in environmental management in UK and its enlightenment to China. *China Population, Resource and Environment* 14 (5): 125–129.

The manufacturer's authorised representative in the EU is Springer Nature Customer Service Centre GmbH, Europaplatz 3, 69115 Heidelberg, Germany. If you have any concerns regarding our products, please contact ProductSafety@springernature.com

Printed and bound by CPI Group (UK) Ltd, Croydon, CR0 4YY
30/09/2026
02225318-0007